T0137336

An Intelligent Inspection Planning System for Prismatic Parts on CMMs

Slavenko M. Stojadinović · Vidosav D. Majstorović

An Intelligent Inspection Planning System for Prismatic Parts on CMMs

 Springer

Slavenko M. Stojadinović
Department of Production Engineering
Faculty of Mechanical Engineering
University of Belgrade
Belgrade, Serbia

Vidosav D. Majstorović
Department of Production Engineering
Faculty of Mechanical Engineering
University of Belgrade
Belgrade, Serbia

ISBN 978-3-030-12809-8 ISBN 978-3-030-12807-4 (eBook)
https://doi.org/10.1007/978-3-030-12807-4

Library of Congress Control Number: 2019930962

This Springer imprint is published by the registered company Springer Nature Switzerland AG
The registered company address is: Gewerbestrasse 11, 6330 Cham, Switzerland

Preface

The book examines an intelligent system for the inspection planning of prismatic parts on coordinate measuring machines (CMMs). The content focuses on four main elements: the engineering ontology, the model of inspection planning for prismatic parts on CMMs, the optimisation model of the measuring path based on an ant colony approach, and the model of probe configuration and setup planning based on a genetic algorithm. The model of inspection planning for CMMs developed here addresses inspection feature construction, the sampling strategy, probe accessibility analysis, automated collision-free operation and probe path planning.

An experiment was performed on two parts that have been manufactured for the purpose of this research. The inspection results show that all tolerances for both parts are within the specified limits. The optimised path is compared with the measuring path obtained by online programming on CMM and with the measuring path obtained in CMM module for inspection in Pro/ENGINEER software. Results of a comparison between the optimised path and the other two generated paths show that the optimised path is at least 20% shorter than the path obtained by online programming on CMM, and at least 10% shorter than the path obtained using CMM module in Pro/ENGINEER.

The proposed model offers a novel approach to intelligent inspection, while also minimising human involvement (and thus the risk of human error) through intelligent planning of the probe configuration and part setup. The advantages of this approach include: reduced preparation times due to the automatic generation of a measuring protocol; potential optimisation of the measuring probe path, i.e. less time needed for the actual measurement; and increased planning process autonomy through minimal human involvement in the setup analysis and probe configuration.

Belgrade, Serbia

Slavenko M. Stojadinović
Vidosav D. Majstorović

Contents

List of Figures

List of Tables

Chapter 1
Introduction and Review of Inspection Planning Methods

Abstract Research and development of intelligent systems for inspection planning on coordinate measuring machines (CMMs) present a precondition for the development of a new generation of technological systems and their application in a digital quality concept, which is based on a global product interoperability model [1–5] where CAD-CAM-CAI information is integrated within a digital platform. This approach presents a basis for virtualisation, simulation and planning inside metrological systems, particularly for the inspection of prismatic parts (PMPs) on a CMM. Research conducted within this book relates to the field of inspection planning for the metrologically complex prismatic parts on a CMM. In a broad sense, the research is directed to the development of the local and global inspection plan for prismatic parts on a CMM. In a narrow sense, it encompasses determination of inspection sequences for metrological features, determination of the number and position of measuring points, as well as the optimal measuring probe path.

1.1 Introduction Remarks

PMPs are an important group of mechanical parts frequently used in industry. PMPs consist of the basic geometric features such as plane, cylinder, cone, etc. From the metrological aspect, this group also implies free-form surfaces whose inspection is not strictly required, and they are present mainly due to esthetical or some related reasons. PMPs are present in almost all types of manufacturing. In this work, our focus is on parts with medium and high quality of dimension tolerances and surfaces roughness precision used to build machine tools.

CMM is a basic element of flexible automation in production metrology and represents an unavoidable factor in prismatic parts inspection. Today, the ongoing intensive research relates to solving the problem of intelligent inspection planning on a CMM, as a prerequisite for the development of intelligent measuring machines, on the one hand, and reduction of measuring time given the increasingly remarkable geometric and functional variability of products, on the other hand.

© Springer Nature Switzerland AG 2019

S. M. Stojadinović and V. D. Majstorović, *An Intelligent Inspection Planning System for Prismatic Parts on CMMs*, https://doi.org/10.1007/978-3-030-12807-4_1

Inspection performance on measurement machines is based on complex software support for various classes of metrological tasks (tolerances). Implementation of uniform inspection planning for such machines is a special problem that depends on the metrological complexity of prismatic parts, inspection planner's intuition and experience-based knowledge. Elimination of intuition, presentation of knowledge, reuse and share of knowledge through the development of intelligent system for inspection planning capable of decision-making at a given moment are solutions of the mentioned problem. In developing the intelligent system, the emphasis is placed on generating the optimal measurement sensor (probe) path as a fundamental segment of prismatic parts inspection. In general, the concept involves:

- development of the ontological knowledge base through the definition of entities and rules on the basis of which it will be searched for the preparation of geometric information and their connection and integration with standard forms of tolerances indirectly, introducing metrological features and their reduction to geometric features,
- development of a global inspection plan that defines an optimal sequence of the metrological features inspection through the analysis of a measuring sensor accessibility and grouping of the features according to the access directions,
- development of a local inspection plan that will generate the number and position of the measuring points, i.e. their distribution for all metrological features and an optimal measuring sensor path for the points thus distributed.

The actuality of planned research is reflected in several research themes related to:

- metrological interoperability,
- intelligent production metrology and
- digital technology systems.

The significance of starting the research and motivation is in the achieved results within the research projects launched in the USA by the National Institute of Standard and Technology (Metrology Project Team) and leading European metrological laboratories, as well as by IMEKO organisation [1]. Research relates to the development of an intelligent concept of prismatic parts inspection planning in order to reduce overall measuring time on a CMM through reduction of the component of time required for inspection planning. All this research aims to organise and effectively use knowledge (experience-based, intuitive) and deduction based on it, and in order to implement a uniform inspection plan given a relatively high metrological complexity and variability of products.

The crucial problem in prismatic parts inspection planning is connecting tolerated measures of prismatic parts with the basic geometric features according to GD&T, PMI, IPIM and bridging the gap between two partially successfully developed approaches, geometric and tolerancing, for the development of the computer-aided inspection process planning (CAIPP) system. One of the first steps in the development of the intelligent concept of inspection is integration of geometry and tolerances of prismatic parts that are involved together in inspection planning. In order to achieve this, it is necessary to develop a knowledge base that answers the question of which geometric feature takes part in creating of which tolerance, i.e. a model of generating the collision-free measuring path based on defined relationship between the geometry and the tolerance provided by a knowledge base. The geometry of the part is thus replaced by a set of points or point-to-point measuring path, the path corresponding to the inspection of the specified tolerances of prismatic parts. Which of the features or group of features will be measured and when are also open problems whose solution yields an optimal measuring sensor path and establishes a relationship between ideal and real geometry of prismatic parts from the metrological aspect. Research conducted in this dissertation represents an attempt to solve previously mentioned two problems in order to realise a long-expected system of intelligent inspection planning for prismatic parts on a CMM.

The notion intelligent can be defined in the literature in two ways. The first implies defining the architecture of intelligent system which, among other things, contains the sensor information as a consequence of the interaction of the system with the environment. Such example was stimulated and developed by the author Albus [2]. The second way, also present in the literature, relates to the application of some of the artificial intelligence (AI) tools in the context of increasing the system's autonomy, optimisation, etc. An intelligent system of inspection planning is reflected in this book in the application of AI tools such as engineering ontology, ant colony and genetic algorithm. Engineering ontology is used for ontological structure modelling in order to define the relationships between tolerated measures (specified tolerances) and geometry of prismatic parts, which participates in the creation of tolerances. The ontological structure is then used as a basis for defining a knowledge base, which answers the question: 'Which geometric feature takes part in creating which tolerance?' The answer is obtained by searching the ontological structure given by a graph of knowledge base using the so-called reasoner after implementation in software Protégé. The intelligent system, on the other hand, also refers to the application of one of the swarm theories, in this case the ant colony for the measuring path optimisation and genetic algorithm for PMPs setup and optimisation of the sensor configuration.

The initial measuring path defined from point-to-point represents the input for the ant colony-based optimisation model. Each ant from the colony visits all node points randomly and lays pheromone along the routes it passes. After a certain number of cycles, which can be specified, the pheromone trails are formed. The pheromone trails which have the largest number of laid pheromone represent the shortest measuring path or optimised measuring path.

A set of possible PMP setups and probe configurations for two types of sensors (probe star and probe head) is reduced to optimal number by GA-based methodology. For each part setup, the optimal probe configuration and optimal point-to-point measuring path are possible to obtain.

Having in mind the above exposed, scope, domain and current research, there are six directions on which conducting the research is based:

- developing a knowledge base model to support an intelligent inspection planning system,
- reducing standard forms of tolerances, through metrological features, to basic geometric features, and thereby establishing direct relationship between the tolerances and the geometric features,
- developing a global inspection plan or method for defining a sequence of basic geometric features inspection,
- extending current methods of the measuring points generation or approximation of the real metrological primitives for variable parameters of basic geometric primitives and taking into account that measuring surfaces of primitives can be internal or external,
- developing an algorithm for the optimal measuring probe path generation by applying the ant colony method,
- developing an analysis of PMP setups and the probe configuration for inspection on a CMM as well as applying the genetic algorithm method.

According to the setup directions and their parallel realisation, scientific contributions of research in this book are the ontological base, developed inspection model and its experimental verification, as well as the optimisation models based on ant colony and genetic algorithm.

The text of the book is organised into six chapters through which the conducted research is presented.

In the introductory chapter, within the analysis of the current state of research and development in the field of inspection planning on measuring machines, essential problems and directions of research and development of inspection planning for prismatic parts on measuring machines are considered from the aspect of intelligent inspection plan definition, highlighting the latest research studies in the world and in this country. Particularly, prominence is given to the problems of metrological interoperability between the metrological software programmes and the problem of a wide gap between GD&T and current standards for definition of CAD geometry, which affects the disruptions in information flow in inspection planning and is responsible, in most cases, for impossibility of involving automatic generation of the relationships between CAD geometry and tolerances as well. Also, a brief review is given of the measuring machines development and inspection planning on them, highlighting the factors that have contributed or caused such chronological order of development of these two segments of coordinate metrology.

It is pointed out, among other things, that market demands should have a quick response to inspection despite high geometric and metrological variability of products, i.e. tendency to the minimum total measuring time through the reduction of the time component needed for inspection planning. On the other hand, it is necessary to minimise the physical and mental involvement of man in certain parts of the measurement process such as inspection planning in order to achieve consistent or uniform quality of inspection, as well as to raise the level of the measurement process autonomy. On the basis of conducted research, key elements of the inspection planning process, their role and relationships between them were determined in order to define the intelligent inspection planning system for prismatic parts.

Chapter 2 commences with the analysis of current state of engineering ontology development in order to define the ontological knowledge base, reuse and distribute knowledge in the domain of coordinate metrology. The primary task of a knowledge base is to connect the geometry of parts with their tolerances using basic ontology components such as classes, individuals and properties. The knowledge base development is preceded by ontology development for the domain of coordinate metrology and methodology of its development at conceptual level. In order to realise this, it was necessary to consider the current methodologies of development and ontologies created to date, as well as to develop and implement own ontology that will serve as a logical structure of a knowledge base for prismatic parts intelligent inspection on CMMs.

Developed ontology was implemented in software Protégé. By defining engineering ontology with the help of the presented method, a set of terms is defined, which mapped into the domain of knowledge base building represents entities and relations between entities. The engineering ontology classes are the knowledge base entities, while relations between entities are the engineering ontology properties. The thus defined ontology was used to define a set of terms of a knowledge base based on a graph and its decision-making based on the so-called reasoner of software Protégé.

Motivation for combining the ontology and a previously developed graph comes from observed suitability of metrological primitives to be represented by means of the basic components of engineering ontology, and by searching the graph, the answer can be obtained to the question of which features a certain standard form of tolerance consists. The response to this question gives the relationship between tolerances and geometry of the part, and the condition for defining a mathematical model of the prismatic parts inspection is further met.

The result of the proposed ontology development method is an iterative process for the coordinate metrology domain in five steps. The result of this approach is also defining the affiliation of geometric features (primitives) to certain forms of tolerances through searching a graph of the knowledge base model. By searching the graph, general forms of tolerances defined by a standard are connected to geometric features so that it is possible to define metrological sequences and plan a measuring sensor path.

In Chap. 3, a developed model of inspection planning for PMPs on a CMM is presented. The main idea of the model is generation of a measuring probe path based on a CAD model for prismatic part and its (specified) tolerances. The model is composed of defining the mathematical model, primitives modelling for inspection, measurement points' distribution, measuring sensor accessibility analysis, collision avoidance and measuring sensor path planning. The key role of the mathematical model is defining the relationships between the measuring machine coordinate systems, measuring part and constituent features of the part, which are involved in creating the prescribed tolerances of the part. Each producer of the software for CMMs has its own developed and implemented mathematical model. The model presented in this book is novel and applicable for the set of prismatic parts inspection.

The next step in model development is primitives modelling for inspection based on basic geometric features and their parameters. The notion geometric features have been first defined in analytical geometry to be applied later, on the same basis, in engineering modelling. In coordinate metrology this notion represents the basis for defining—modelling of features for inspection from the aspect of geometry and tolerances. In this modelling, the geometric features are represented by the lowest level of tolerance definition or the object of generating the measuring sensor points on a measuring part. Each geometric feature is unambiguously determined by a local coordinate system and a set of parameters with respect to it. Geometric features involved in modelling are point, plane, circle, hemisphere, cylinder, cone, truncated cone and truncated hemisphere.

After the definition of features, the next step is measurement points' distribution for thus defined features. The measurement points' distribution is based on the Hammersley sequence for calculating the measuring points coordinates along two axes of a metrological feature. By modifying the Hammersley sequence, the formulas were derived for distribution of the desired number of measuring points for all above-mentioned geometric features. Experimental example involved the distribution of ten measuring points; however, the distribution method allows loading of any number of measuring points, depending on the type and quality of prescribed (required) tolerance.

To perform the collision-free inspection of a single feature, it is necessary to carry out an analysis of the measuring sensor accessibility. It includes determination of another two sets of points and definition of the fullness vector of a primitive, with which it is taken into account whether the feature is full or empty. The first set of points represents the points for the transition from fast to slow speed, whereas the second set of points defines the initiation of fast speed. Using this approach to defining the sets of points and feeds, collision avoidance is performed between a single features and a measuring probe. For the definition of a measuring point and observed three characteristic cases of the surfaces of features, the principles of these two sets of points' distribution are defined. Based on that principle, a formula was derived for the probe path length or initially travelled path, which is a basis for path optimisation by applying ant colony in the next chapter.

When it comes to collision avoidance, it is necessary to avoid a collision in sensor transition from one feature to another (a simple example is parallelism inspection of two planes). For that reason, the principle of collision avoidance was developed based on the STL model of representing the prismatic part geometry, its tolerances, end point coordinates for the preceding feature inspection and start point coordinates for the next feature inspection. The principle consists of the iterative shifting of a straight line, which originally passes through mentioned points, for some value according to a specific procedure, until the moment when the straight line does not intersect with the prismatic part volume. The line is shifted by translation for a given value. The problem of finding the crossing points between the straight line and the prismatic part value is reduced to the problem of finding the crossing point between the surface limited by a triangle and a line segment.

Based on determined relationship between tolerances and geometry, as previously explained procedures and developed methods and principles, a measuring sensor path planning produces as the output a point-to-point measuring path for a given prismatic part.

In Chap. 4, the analysis of primitives was performed from the viewpoint of automatic generation of the measuring probes configuration and PMP setup. The analysis begins from the primitive parameters such as vector of a primitive and fullness vector of a primitive. At the output, the model generates optimal number of the part setups and measuring probes configuration for each of the setups. Optimisation is done using GA technique which has Boolean matrix for the input and which represents the output of the analysis model of PMP setup and measuring probes configuration.

In Chap. 5, the measuring sensor path optimisation is presented during prismatic parts inspection on a CMM. Optimisation model is based on the mathematical model of a developed inspection planning model, which defines the initial path represented as a set of points determined by a sequence for collision-free probe pass and solution of a travelling salesman problem (TSP) applying the ant colony. The chapter consists of several sections such as a data model, a mathematical model for optimisation and an ant colony optimisation (ACO) model based on ant colony.

The basis of ACO model is data taken over from IGES file of the prismatic part CAD model and mathematical model which, at the output, gives a set of points among which the shortest distance or optimal measuring probe path is sought. In defining the initial path by strict determination of the point sequence, a measuring path was defined. In order to provide the space for optimisation and obtain the optimal path, three zones were defined: zone of possible collision, optimal zone and zone of unprofitable inspection planning.

The sets of points which belong to the optimisation zone are allowed accidental visit points or points through the measuring sensor pass during inspection. It is this fact or accommodation of the mathematical model that allows for application of the ant colony-based optimisation model. Namely, the principle of this technique is random tour of all, in this case, points and depositing pheromones along such trail and then finding the shortest path according to the criterion of the largest amount of pheromones along the travelled path. The path that contains the largest amount of pheromones is the shortest path. The number of random tour cycles and the number of ants in a colony can be chosen.

The optimal path is then compared with online programmed path on a measuring machine ZEISS UMM500 and the path generated in Pro/ENGINEER for identical parameters. The results for comparison of optimised path and online programmed path indicate at least 20% lower value of optimised path length, whereas, compared to the Pro/ENGINEER path, at least 10% lower value of optimised path length for identical parameters setting.

In Chap. 6, experimental verification of the developed inspection planning model was carried out and concluding remarks are presented. For path verification and simulation in order to visually inspect collision between the measuring sensor and the workpiece, first the programme was written in appropriate software environment. Apart from verification and simulation, one of the major goals of the written programme is also the generation of a measuring protocol and a control data list at the output, which are then used in the experimental planning process and as the input for experimental measurements. Simulation was developed using three algorithms as follows:

- algorithm for measurement points distribution,
- algorithm for collision avoidance and
- algorithm for measuring probe path planning.

The algorithms include all geometric features represented by the developed inspection planning model. The developed model of measuring sensor path simulation gives at the output a measuring protocol and a control data list for test measuring parts. The measuring protocol is further used for online programming of a measuring machine ZEISS UMM500, programme generation for this machine, measurement process and verification of the developed inspection planning model. For experimental needs, two workpieces were designed and machined.

Machining was performed at the Institute for Machine Tools, Chair of Production Engineering, Faculty of Mechanical Engineering, Belgrade University, on a machining centre HMC500 with surface finish quality N7 and N8.

Prior to the inspection process on designed and machined test workpieces, dimensional accuracy of the machine was checked based on the artefact of domestic production according to standard ISO 10360-2. Part inspection was carried out at Vienna Technical University, in High Precision Measurement Room—Nanometrology Laboratory during visit to the university.

Results of inspection indicate that all tolerances of workpieces are within the limits prescribed by a technical drawing. This confirms that the developed inspection planning model is a successful and practical applicable approach for the development of intelligent inspection planning concept for prismatic parts.

1.2 Review of Inspection Planning Methods

The subject of this section is a review of performed research and results achieved to date in the field of inspection planning on a CMM, highlighting the latest research studies in this country and in the world.

As it mentioned, research projects launched in the USA by the National Institute for Standards and Technology (NIST) and leading European metrological centres are also related, among other things, to the research on development of the intelligent inspection planning system for prismatic parts with the aim of reducing total measuring time on a CMM through reducing the time component required for inspection planning. The aim of all research is to organise and effectively use knowledge (experience-based, intuitive) and deduction based on it and all this for conducting a uniform inspection plan given relatively high both metrological and geometric complexity and variability of products. In addition to mentioned research, other directions of studies are topical such as nanometrology [6–8].

1.2.1 Model of a CMM-Based Measuring System

The starting point that can be used for developing the intelligent inspection planning concept is a model of some current measuring system. Given that in this case it is about a branch of metrology referred to as coordinate metrology and measuring of PMPs on a CMM, such system is CMM-based.

According to the author Zhao, IDEF0 model of dimensional measuring system is composed of four major elements such as:

- defining a product (A1),
- defining a measurement process (A2),
- measurement process (A3), as well as the
- analysis and report on quality data (A0) (Fig. 1.1),

whereas IDEF0 inspection plan model defines the activities of a measuring plan (Fig. 1.2) such as measuring area (A21), measuring method and features (A22), macro-measuring plan (A23), reactive plan and auxiliary data (A24) and micro-measuring plan (A25) in order to ensure the functionality of the part during and after the machining process.

IDEF0 model of measuring process planning for parts manufactured on a machining centre is presented in [9].

1.2.2 A Brief Review of CMMs Development

The development of a CMM is conditioned by the introduction of new technologies and development of technological systems, and, in general, it can be divided into the development of hardware and software for a CMM. According to [10], at the beginning of industrial development, the task of manufacturing, i.e. inspection, was to ensure the product's function with high product tolerances. However, with increase of accuracy, complexity and flexibility of machined parts, establishment of the serial production concept, introduction of the principle of parts interchangeability, transfer and flexible lines, the role of inspection on a CMM does imply not only identification of geometric accuracy but also the machining process control

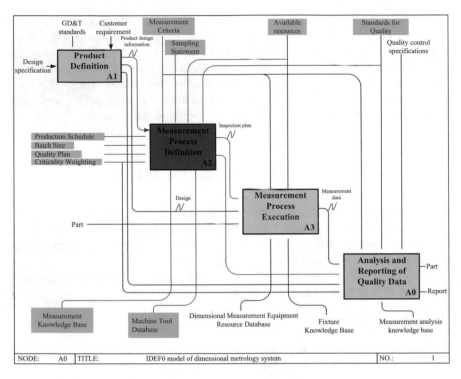

Fig. 1.1 IDEF0 model of dimensional metrological system [3]

parameters such as tool tuning, machining mode parameters—machined surface roughness and the like. According to the same authors, two events were of crucial importance for the development of a CMM:

- beginning of industrial production by the end of 1960s and electronics-based length and angle measuring systems and
- development of computer technology.

Additionally, the development of the science of materials, especially composites, has greatly contributed to the hardware structure development, whereas the software solutions for CMM have been enhanced owing to new error compensation models. Namely, unlike the conventional production concept that relied upon tendency to building CMM components as accurately as possible, which considerably increased manufacturing costs, today a new concept is in effect, where required CMM accuracy is accomplished by algorithms for error compensation and accuracy in building measuring machines' components.

One concept of modern machines development defined at the end of twentieth century is presented in [11] and state of the art in production metrology development by the middle of the previous decade in [12]. The significance of production metrology in industry is described in [13].

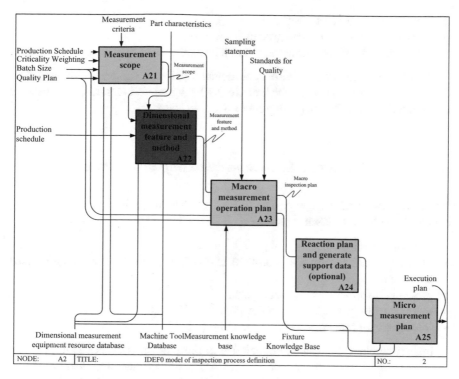

Fig. 1.2 IDEF0 diagram of the measuring process [3]

1.2.2.1   Software Systems for CMM

In parallel with hardware development for CMM, there was the development of software systems which can be classified into two groups:

- general purpose software and
- special purpose software (e.g. inspection of gears and turbine blades).

According to the complexity level of metrological tasks [10], software systems for CMM can be arranged into the following groups:

- measuring and inspection of length tolerances, angularity and location tolerance,
- measuring and inspection of gear parameters,
- measuring and inspection of curves and surfaces.

Basic geometric elements that are used to accomplish different metrological tasks, with the help of software for CMM, are point, straight line, circle, plane, cylinder, cone, sphere and torus. As it is known, for each of them, the minimum number and maximum number of points are determined, by means of which, using the method of least squares, it is unambiguously defined.

Software can present measurement results in different types of coordinate systems such as:

- the Cartesian coordinate system (x, y, z),
- polar-cylindrical coordinate system (r, θ, z) and
- spherical (r, θ, y) measuring coordinate system related to a measuring part.

Software development is also conditioned by the development of statistical theories and algorithms [14], which are in the background, i.e. used for processing of data measured in the inspection process on a CMM.

1.2.3   Development of Inspection Planning on a CMM

The development of inspection planning on CMMs in the last three decades has passed through the following phases:

- manual planning,
- planning generated by CAI software which is still the most commonly used approach,
- planning obtained by an expert system [15],
- automatically generated measuring plan [16–21] and
- intelligent concept of inspection planning [22–30].

Inspection plans that cannot be included in those mentioned above are [31, 32].

Over the past years, models of virtual CMMs are designed and developed [33, 34], particularly for the needs of measurement uncertainty modelling [35, 36].

If man is excluded as a planner and developer of the manual inspection plan, as well as the original software for CMM, a model shown in Fig. 1.3 belongs to the earliest knowledge-based inspection planning systems. It contained all module-form elements of inspection planning and something that occurs for the first time, and it is the element for decision-making referred to as an expert tolerancing consultant.

One nearly comprehensive importance of production metrology development is presented in [37], whereas the object-oriented approach to inspection planning is given in [38]. Generation of the standardised reference data set for measuring machine was performed in [39].

Two major parameters dictating such sequence of the inspection planning approaches development are achievable machining accuracy and total measuring time (Fig. 1.4).

The tendency to reducing total measuring time and increasing of machining accuracy, with emergence of high product variability, requires steady development of new inspection planning methods on a CMM. The common element that has remained the same throughout all these development stages is a measuring part, i.e. the object of contact—measuring.

According to the manner of analysis and synthesis of workpiece geometric information (tolerances), three approaches could be distinguished:

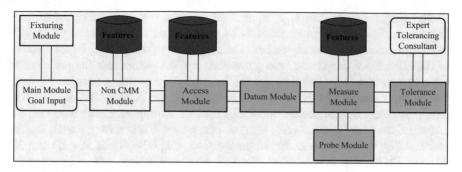

Fig. 1.3 Inspection planning model [15]

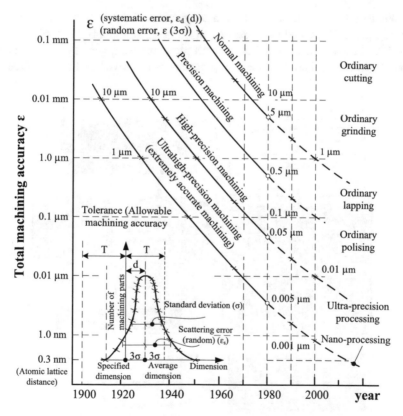

Fig. 1.4 Prediction of achievable machining accuracy according to Taniguchi [40]

- analysis of geometry [41–43],
- analysis of tolerances [44–49], and
- combined approach [50–52].

The inspection process is composed of several key elements such as path planning, collision avoidance, accessibility analysis and a workpiece setup, as well as configuration of measuring probes. The level of analysis and implementation of to date developed inspection plan generation models determines the presence of these elements in it. A complete system for inspection planning contains all mentioned key elements. In the work by Hwang [53], the elements present are part setup, determination of probes orientation and measuring sequence based on the features library for inspection (Fig. 1.5). A purpose of Weckenmann's analyses, the practicability of extensive analytical comparison tests of different 1D, 2D and 3D artefacts [21] is the determination of meaningful positioning of the artefact in the measuring volume of the machine.

Models of information necessary for inspection planning are given in [39, 55, 56].

Note that the measuring strategy also plays an important role in inspection planning [57, 58], as well as measuring machines programming [59, 60].

In [54, 61–64], the approaches for path planning are given. Typical inspection planning procedure based on these approaches is displayed in Fig. 1.6. One step

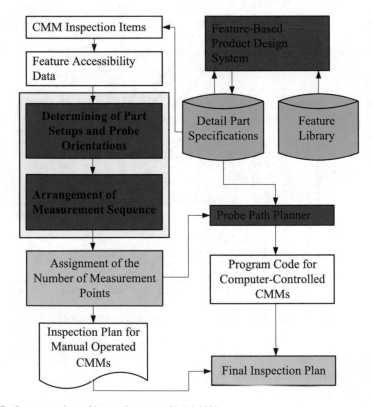

Fig. 1.5 One operation of inspection on a CMM [53]

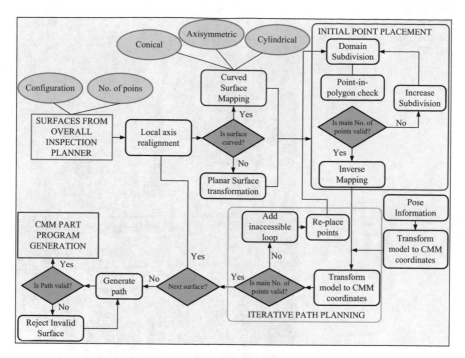

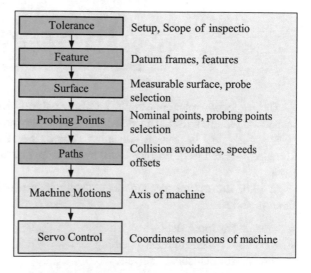

Fig. 1.6 Inspection planner's algorithm on a CMM [54]

Fig. 1.7 One control
inspection hierarchy [65]

Tolerance	Setup, Scope of inspectio
Feature	Datum frames, features
Surface	Measurable surface, probe selection
Probing Points	Nominal points, probing points selection
Paths	Collision avoidance, speeds offsets
Machine Motions	Axis of machine
Servo Control	Coordinates motions of machine

forward was made by Juster who, besides the measuring path, defines the control
code for CMM based on a hierarchy given in Fig. 1.7.

The works [66, 67] consider solutions for collision avoidance, whereas [68–76]
the accessibility analysis and [76–81] the measuring probes configuration.

Fig. 1.8 Scheme of 'on-machining' measurements [96, 97]

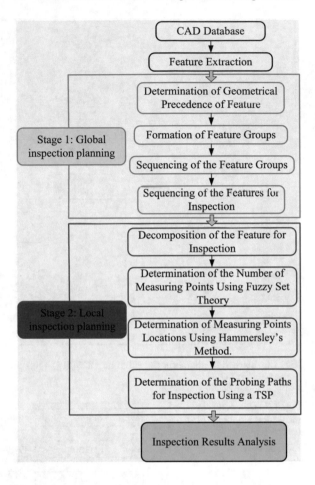

The inspection planning could be also analysed through the local and global inspection planning [82]. The elements of these two plans are presented in Fig. 1.8.

Systems for inspection planning could be feature-based [83–92] and knowledge-based [93]. Feature-based systems are also used for machining planning [94].

In [95], the offline measurement planning system is presented, composed of three modules:

- module for input data,
- module for measurement planning and
- module for statistical analysis.

An example of the feature extraction from a CAD model for the needs of inspection planning, based on CBR technique, is shown in Fig. 1.9. This study is the first generative approach to inspection planning by applying CBR technique and considers only the geometric aspect of a measuring part but not its tolerances.

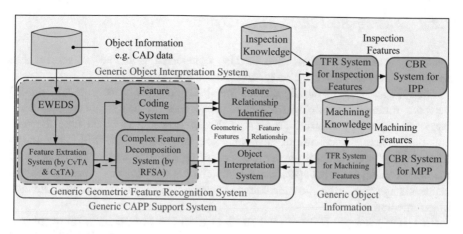

Fig. 1.9 Framework of inspection planning according to [98, 99]

The state of research in the field of inspection planning for prismatic parts and sculpture surfaces is given in detail in [82, 100].

Automatic inspection planning on a CMM reported in [101] is based on boundary representation (B-representation) of a solid model in SAT format. At the output, it gives the programme for CMM in DMIS format.

In coordinate metrology, such as measurement of PMPs on a CMM, the definition of the workpiece from the geometric aspect and the aspect of tolerances should be considered together. From the geometric aspect, the workpiece is characterised by geometric features (e.g. plane, cylinder, sphere), but from the aspect of tolerances, the workpiece is characterised by metrological features (e.g. distance between two planes or two cylinders). The link between these two types of features (geometric and metrological) implies the type of tolerance. Based on this, the measuring probe path accordingly could be considered as a set of points composed of three subsets:

- The first subset contains measuring points that can be obtained based on geometric information.
- The second subset contains the remaining points that measuring probe uses for the inspection of one geometric feature that could be also obtained from geometric information.
- The third subset enables to avoid a collision, and this set must be defined based on tolerance information (i.e. link between two geometric features) and the collision avoidance principle in the transition of a probe from one geometric feature to another.

1.2.4   Metrological Interoperability

Inspection planning for PMP can be also considered from the viewpoint of a relatively new paradigm such as metrological interoperability.

Metrological interoperability implies the ability for joint work (communication) between different metrological systems, techniques or programmes (software), based on different platforms. To arrange this, it is necessary to develop the procedures, enhance the current standards or interfaces that will link them. Metrological interoperability is defined, according to [102], as the specification of the interface language that specifies and implements metrological data so that users can exchange them with the manufacturer (vendor) using minimum but quite sufficient amount of data.

In [103], interoperability is referred to as a measure of ability of efficient and precise communication between several systems, whereby solution of different tasks is provided. The major challenge of metrological interoperability is determination of appropriate procedures and interfaces based on a joint language, so that manufacturers are supplied with all necessary functions, having enough space for their products to make them different from their competitors' products without disclosure of business secrets. The goal of metrological interoperability is the development and maintenance of joint languages for key interfaces in big industrial organisations (e.g. AIAG, I++, BOEING) that will be widely accepted.

A common goal achievement involves users (Daimler Chrysler, Ford, BMW, etc.), vendors (Zeiss, Mitutoyo, etc.), industrial organisations, as well as standardisation and accreditation bodies.

An example of interoperability in the domain of dimensional metrology is given in Fig. 1.10. It is composed of four general activities:

- design,
- planning process,
- execution process and
- analysis.

Mentioned activities in the dimensional metrology domain are matched by CAD-PMI activities, measurement process planning, measuring (measuring equipment), results and analysis, respectively.

1.2.4.1   Problem of Metrological Interoperability

Interoperability problems are permanent, occurring in applications development, especially in their usage. Today's applications miss unambiguous description or definition at a single feature level, and deterministic definition of relationships

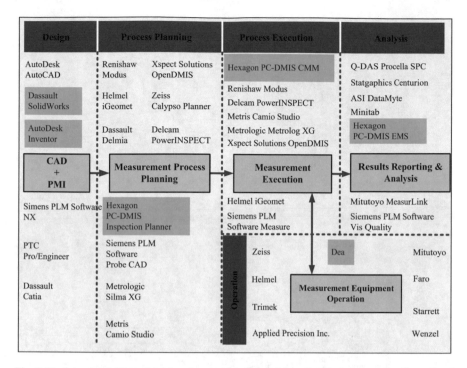

Fig. 1.10 Interoperability obstacles between commercial software systems in dimensional metrology [3]

towards higher levels. One approach to solving this problem is similar to today's everyday language based on semantic description of the problem. Misunderstanding between people during business operations within a production system and ad hoc translations of software applications contributes to the costs rise of interoperability in production, which is not desirable.

Interoperability issues occur in production plants when it is desired to establish a balance between running (inspection) of several measuring machines by inspection displacement from one machine to another or when it is necessary to transfer the inspection programme for one part or group of parts from one plant to another. One example of the current state of the interoperability problem is presented in Fig. 1.11.

Interoperability problem can be solved by developing a new language that will be used by all (manufacturers and users). A few years ago, NIST initiated metrological interoperability of products between the hardware and software manufacturers. In [104], a complete survey of tests is given for identifying the achieved interoperability level of software and hardware developed by NIST.

One example of the interoperability problem solution in the domain of dimensional metrology is given in Fig. 1.12, while the solution by applying engineering ontology is reported in [105]. The problem of metrological interoperability is discussed in more detail in [106, 107].

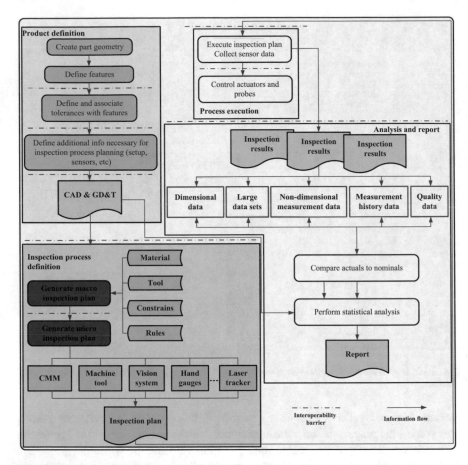

Fig. 1.11 Current state of interoperability problem in dimensional metrology [3]

Despite great efforts, interoperability failure is ever-present and represents a severe obstacle to productivity growth and efficiency in industry. Through a survey of to date achieved results in the sphere of metrological interoperability, the present book highlights the significance and contribution of the intelligent inspection planning system as one of the methods for eliminating interoperability obstacles, i.e. solution of the interoperability problem in the domain of inspection for PMP on a CMM.

1.2.4.2 Project Team for Metrological Interoperability

NIST was founded by Automotive Industry Action Group (AIAG), which consists of several corporations and universities, and their collaboration is a basis for solving problems in the area of metrological systems.

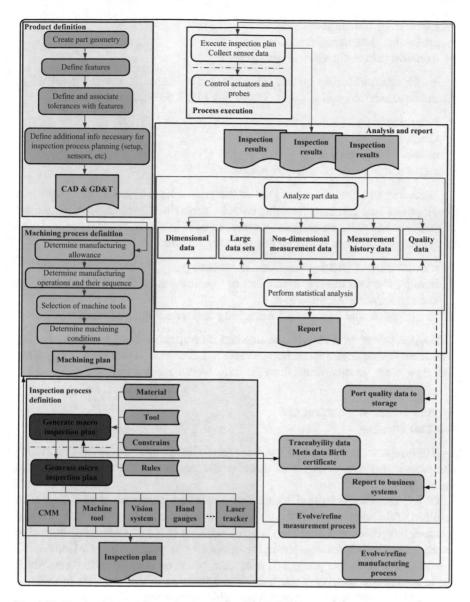

Fig. 1.12 Future vision of dimensional metrological system [3]

AIAG group is a founder of the project team for metrological interoperability, so-called Metrology Project Team (MPT). Apart from the problems of metrological interoperability, as a primary problem, the team deals with other fields such as:

- production quality data,
- production data metrics,

- engineering visualisation,
- production data management and
- technical data exchange and management.

As for the measuring system type, the project team is mainly involved in dimensional metrological systems, whose key interface components are:

- Programming,
- CAD,
- measuring experiment,
- report and analysis of obtained results.

According to [108], the basic goal of the project team is to reduce the product development time and manufacturing costs by achieving high interoperability level of hardware and software components that are used in automated metrology. Some of the activities underway within this project team are:

- identification of existing standards shortcomings,
- introduction into in-depth evaluation of existing standards and development of standards for certain interfaces,
- identification and support in harmonising and overlapping of standards.

Besides, this team deals with the standard infrastructure development, as well as with harmonising standards of Europe and the USA. The project team is organised into three work groups whose thematic frameworks are as follows:

- programming,
- joint format of the report and
- CAD interface.

STEP represents ISO support to interoperability problem solution by defining computer-interpretative data that describe the product and the method of data exchange.

DMIS describes a neutral format for inspection programmes and inspection results [109], whereas I++DME describes a communication interface between measuring machine and execution control software [110] and represents a road to interoperability problem solution in the domain of inspection on a CMM.

AIAG also promotes product development based on standard interfaces, which are in accordance with open standards. The advantage is that products of different manufacturers will be interconnected and without conversion of one data format to another. The benefits for product users supported by standard interfaces are:

- ease of products connectivity,
- reduction of maintenance costs,
- flexibility in users' choice of components and
- increased competition among vendors, followed by costs reduction.

1.3 Formulation of the Research Problem

When creating a measuring protocol, the geometry of a measuring part is taken into account or loaded in the software for CMM programming in the form of some output file such as neutral data format, IGES or STEP, while data on tolerances are loaded from a technical drawing. For the time being, there are no data or file format, where data on both geometry and tolerances are stored. Some CAD software systems such as Autodesk Inventor Professional 2011 have the possibility of specifying, in 3D environment, some forms of tolerances like tolerances of length and diameter; however, data format with visible relationship between specified tolerances and part geometry involved in creating given tolerance does not exist today.

The present-day STEP data model does not have enough GD&T information for automatic generation of the inspection process plan [3]. Figure 1.13 shows EXPRESS-G diagram of proposed GD&T definitions in proposed Application Reference Model (ARM). STEP-NC contains in itself tolerances, but not data to connecting them to the geometry of a measuring part. From geometric viewpoint, an important role is played by the development of GPS area [111].

Clear determination of these data would facilitate automatic inspection based on CAD model of the part and eliminate use of a technical drawing like terms in the figure that are defined by GD&T standard of basic medium on tolerancing information.

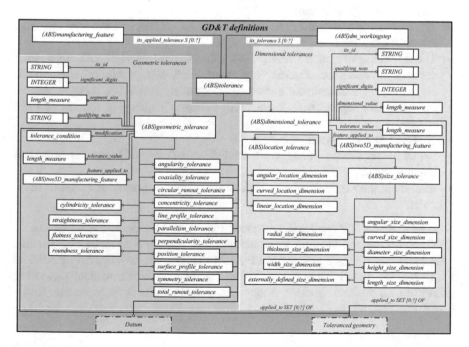

Fig. 1.13 Defining tolerancing information according to GD&T [3]

According to Zhao, the major problems occurring in the process of product definition and measuring are:

- GD&T data must be modelled inside CAD data, but not provided as comments or records. ISO 10303 AP 203 [112] is the only developed standard of data representation supported by all CAD systems; however, it does not contain in itself a tolerances model, components such as a coordinate system of features and tolerances. ISO 10303 AP 224 [113] (feature-based representation) incorporates a model of tolerancing features but as such is not supported by CAD systems.
- Currently, one of the major supports to standard development is a new version of ISO 10303 AP 203 which is modelling tolerance elements. The latest testing conducted by the major CAD suppliers was applied to the interpretations of GD&T information modelled in AP 203 in second edition [114]. GD&T definitions from AP 214 (data core for automatic design of machine processes) are harmonised in AP 203 in second edition. These GD&T definitions are mainly applicable for interpretation purposes, and therefore, they are not sufficient for automatic generation of the measurement process plan. Further harmonisation of GD&T definitions is needed between AP 214 and AP 224, as well as their adoption in AP 203. It is only in this way that AP 203 will be able to provide adequate information for generation of the measurement process plans.
- Non-standard GD&T information is related to the design of part geometry. This fact pervades both the process of product definition and definition of the measurement process activities.
- The absence of standard mechanisms to also encompass one standard language for measuring methods, practice and rules.
- The absence of computer readable (interpretative) and standard definitions of the capacity resources for measuring equipment, capacities, possible configurations, performances, measurement uncertainties, sensors, auxiliaries, rotary tables.
- Poor end-user support for non-exclusive interface language of the metrological system.
- I++DME is not a formal standard.
- I++DME application must be extended to other equipment, sensors and environments.
- Implementation barriers of I++DME must be reduced, such as initial or input costs.

In proposed STEP application protocol, the ARM data model for integration of CNC machining and inspection and definition of dimensional measuring features were harmonised by AP 219 and HIPP ARM data models as proposed by NIST at 53rd ISO TC 184 SC meeting in Dallas in 2007. One-dimensional measuring primitive (dm_feature) is either simple (dm_simple_feature) or composite (dm_-composite_feature), as shown in Fig. 1.14.

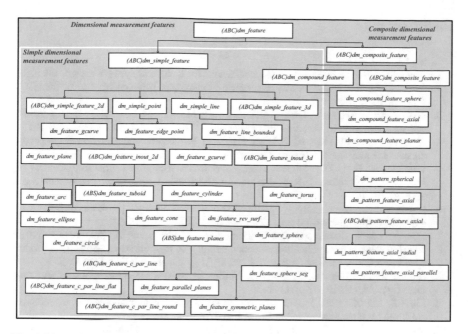

Fig. 1.14 EXPRESS-G diagram of metrological primitives [3]

Each simple or composite primitive can be one of the several types that identify several joint primitives in the measurement process planning—such as cones, cylinders, and composite primitives like patterns (recursive elements).

The above-mentioned problems directly affect interoperability realisation and interfere with the development of automatic and intelligent concept of inspection on a CMM. The attempt to point to the relevance of solution to these problems is presented in [115].

The solution of mentioned problem can be observed in the measuring part decomposition into geometric features and determination of their affiliation to a certain form of tolerance. Furthermore, geometric features can be represented as engineering ontology classes, a novel technique of AI. Interconnection of tolerances and geometric features can be represented through properties of classes and individuals in a unique ontological database. The basic terms in the figure are defined by the EXPRESS formal language. The task of database is to give metrological primitives at the output, i.e. relationships between geometric primitives and tolerances of a measuring part.

The exposed approach to the solution also requires definition of engineering ontology in the domain of prismatic parts inspection on a CMM, which is a subject of the next chapter.

References

1. http://www.imeko.org/
2. Albus SJ (1993) A reference model architecture for intelligent systems design. Intelligent Systems Division, Manufacturing Engineering Laboratory, National Institute of Standards, Technology, Gaithersburg
3. Zhao Y, Xu X, Kramer T, Proctor F, Horst J (2011) Dimensional metrology interoperability and standardization in manufacturing systems. Comput Stand Interfaces 33(6):541–555
4. Zhao F, Brown JR, Kramer RT, Xu X (2011) Information modeling for interoperable dimensional metrology. Springer, London
5. Westkamper E (2007) Digital manufacturing in the global era. In: Cunha PF, Maropoulos PG (eds) Digital enterprise technology, perspectives, future challenges. Springer, Stuttgart
6. Durakbasa N, Bauer J, Bas G (2012) Developments in high precision metrology for advanced manufacturing. In: Proceedings of the ICMEM, pp 210–215
7. Durakbasa MN, Herbert PO, Bas G, Demircioglu P, Cakmakci M, Hornikova A (2012) Novel developments in dimensional nanometrology in the context of geometrical product specifications and verification (GPS). J Autom Mob Robot Intell Syst 6(2):22–25
8. Hansen NH, Carneiro K, Haitjema H, Chiffre DL (2006) Dimensional micro and nano metrology. Ann CIRP 55(2):721–743
9. Lin CZ, Lin ZC (2000) IDEF0 model of the measurement planning for a workpiece machined by a machining centre. Precis Eng 16:656–667
10. Majstorovic V, Hodolic J (1998) Coordinate measuring machine. Faculty of Technical Science, Novi Sad, Serbia. ISBN 86-499-0091-7
11. Wu X, Zhang G (1997) Development of a modern co-ordinate measuring machine. Nanotechnol Precis Eng 114:186–190
12. Weckenmann A, Kraemer P, Hoffmann J (2007) Manufacturing metrology—state of the art and prospects. In: Proceedings of the 9th international symposium on measurement and quality control, Manufacturing Engineering Section, Department of Mechanical Engineering, Indian Institute of Technology Madras, Chennai, India, pp 21–27
13. Durakbasa MN, Osanna HP (2009) Quality in industry. Vienna University of Technology, Wien, Austria
14. Lubke K, Sun Z, Goch G (2012) Three-dimensional holistic approximation of measured points combined with an automatic separation algorithm. CIRP Ann Manuf Technol 61:499–502
15. ElMaraghy HA, Gu PH (1987) Expert system for inspection planning. Ann CIRP 36(1):85–89
16. Limaiem A, ElMaraghy AH (1998) Automatic path planning for coordinate measuring machine. In: Proceedings of the 1998 IEEE, international conference on robotics and automation, Leuven, Belgium, pp 887–892
17. Zhao H, Kruth JP, Gestel NV, Boeckmans B, Bleys P (2012) Automated dimensional inspection planning using the combination of laser scanner and tactile probe. Measurement 45:1057–1066
18. Ravishankar S, Dutt HNV, Gurumoorthy B (2010) Automated inspection of aircraft parts using a modified ICP algorithm. Int J Adv Manuf Technol 46:227–236
19. Chang HC, Lin AC (2010) Automatic inspection of turbine blades using 5-axis coordinate measurement machine. Int J Comput Integr Manuf 23(12):1071–1081
20. Chang HC, Lin AC (2011) Five-axis automated measurement by coordinate measuring machine. Int J Adv Manuf Technol 55:657–673

21. Yau HT, Menq CH (2005) Automated CMM path planning for dimensional inspection of dies and molds having complex surface. Int J Mach Tools Manuf 35(6):861–876
22. Lu CG, Morton D, Myler P, Wu MH (1995) An artificial intelligent (AI) inspection path management for multiple tasks measurement on coordinate measuring machine (CMM): an application of neural network technology. In: Proceedings of the 95 engineering management conference, IEEE, Singapore, pp 353–357
23. Yau TH (1991) The development of an intelligent dimensional inspection environment using coordinate measuring machines. Doctoral dissertation, The Ohio State University, Columbus
24. Lu CG, Morton D, Wu MH, Myler P (1999) Genetic algorithm modelling and solution of inspection path planning on a coordinate measuring machine (CMM). Int J Adv Manuf Technol 15:409–416
25. Stojadinovic S, Majstorović V (2012) Towards the development of feature-based ontology for inspection planning system on CMM. J Mach Eng 12(1):89–98
26. Liangsheng Q, Guanhua X, Guohua W (1998) Optimization of the measuring path on a coordinate measuring machine using genetic algorithms. Measurement 22:159–170
27. Roy U, Xu Y, Wang L (1994) Development of an intelligent inspection planning system in an object oriented programming environment. Comput Integr Manuf Syst 7(4):240–246
28. Zhang GX, Liu SG, Ma XH, Wang JL, Wu YQ, Li Z (2002) Towards the intelligent CMM. Ann CIRP 51(1):437–442
29. Osanna PH (1997) Intelligent production metrology—a powerful tool for intelligent manufacturing. Elektrotechnik Informationstechnik 4(114):162–168
30. Kuang CF, Ming CL (1998) Intelligent planning of CAD—directed inspection for coordinate measuring machine. Comput Integr Manuf Syst 11(1–2):43–51
31. Gerhardt LA, Hyun K (1995) View planning applied to coordinate measuring machine (CMM) measurement. In: Proceedings IEEE conference on industrial automation and control emerging technology applications, IEEE, Taipei, pp 540–544
32. Spitz NS, Requicha GAA (1999) Hierarchical constraint satisfaction for high-level dimensional inspection planning. In: Proceedings of the 1999 IEEE, international symposium on assembly and task planning, IEEE, Porto, pp 374–380
33. Hu Y, Yang Q, Sun X (2012) Design, implementation, and testing of advanced virtual coordinate-measuring machines. IEEE Trans Instrum Meas 5(61):1368–1376
34. Sładek J, Gąska A, Olszewska M, Kupiec R, Krawczyk M (2013) Virtual coordinate measuring machine built using lasertracer system and spherical standard. Metrol Meas Syst 20(1):77–86
35. Sładek J, Gaska A (2012) Evaluation of coordinate measurement uncertainty with use of virtual machine model based on Monte Carlo method. Measurement 45(1):1564–1575
36. Aggogeri F, Barbato G, Barini ME, Genta G, Levi R (2011) Measurement uncertainty assessment of coordinate measuring machines by simulation and planned experimentation. CIRP J Manuf Sci Technol 4:51–56
37. Kunzmann H, Pfeifer T, Schmitt R, Schwenke H, Weckenmann A (2005) Productive metrology—adding value to manufacture. Ann CIRP 54(2):155–168
38. Beg J, Shunmugam S (2002) An object oriented planner for inspection of prismatic parts—OOPIPP. Int J Adv Manuf Technol 19:905–916
39. Lin J-Y, Damodharan K, Shakarji C (2001) Standardised reference data sets generation for coordinate measuring machine (CMM) software assessment. Int J Adv Manuf Technol 18:819–830
40. Taniguchi N (1983) Current status in, and future trends of, ultra precision machining and ultrafine materials processing. Ann CIRP 32(2):573–582
41. Zhao X, Kethara PMT, Wilhelm GR (2006) Modeling and representation of geometric tolerances information in integrated measurement processes. Comput Ind 57(4):319–330

42. Stefano DP, Bianconi F, Angelo DL (2004) An approach for feature semantics recognition in geometric models. Comput Aided Des 36:993–1009
43. Lin J-Y, Mahabaleshwarkar R, Massina E (2001) CAD-based CMM dimensional inspection path planning—a generic algorithm. Robotica 19:137–148
44. Zhou F, Kuo CT, Huang HS, Zhang IIC (2002) Form feature and tolerance transfer from a 3D model to a set-up planning system. Int J Adv Manuf Technol 19:88–96
45. Jiang CB, Chiu DS (2002) Form tolerance—based measurement point determination with CMM. J Intell Manuf 13:101–108
46. Salomons OW, Poerink Jonge HJ, Haalboom FJ, Slooten F, Houten FJAM, Kals HJJ (1996) A computer aided tolerancing tool I: tolerance specification. Comput Ind 31:161–174
47. Salomons OW, Haalboom FJ, Poerink Jonge HJ, Slooten F, Van Houten FJAM (1996) A computer aided tolerancing tool II: tolerance analysis. Int J Comput Integr Manuf 31:175–186
48. Mohib NMA, ElMaraghy AH (2010) Tolerance-based localization algorithm: form tolerance verification application. Int J Adv Manuf Technol 47:581–595
49. Ge Q, Chen B, Smith P, Menq HC (1992) Tolerance specification and comparative analysis for computer-integrated dimensional inspection. Int J Prod Res 30(9):2173–2197
50. Cho MW, Seo TI (2002) inspection planning strategy for the on-machine measurement process based on CAD/CAM/CAI integration. Int J Adv Manuf Technol 19:607–617
51. Bhaskar Sathi VS, Rao PVM (2009) STEP to DMIS: automated generation of inspection plans from cad data. In: Proceedings of the 5th annual IEEE conference on automation science and engineering, Bangalore, India, August 22–25, pp 519–524
52. Zhao F, Xu X, Xie S (1998) STEP-NC enabled online inspection in support of closed-loop machining. Robot Comput-Integr Manuf 24:200–216
53. Hwang YC, Tsai YC, Chang AC (2004) Efficient inspection planning for coordinate measuring machines. Int J Adv Manuf Technol 23:732–742
54. Albuquerque AV, Liou WF, Mitchell RO (2000) Inspection point placement and path planning algorithms for automatic CMM inspection. Int J Comput Integr Manuf 13(2):107–120
55. Barreiro J, Martinez S, Labarga JE, Cuesta E (2005) Validation of an information model for inspection with CMM. Int J Mach Tools Manuf 45:819–829
56. McCaleb RM (1999) A conceptual data model of datum systems. J Res Nat Inst Stand Technol 104(4):349–400
57. Vogel M, Ebinger N, Rosenberger M, Lin L (2010) A novel strategy for cost-efficient measurements with coordinate measurement machines. J Phys 238:1–6
58. Flack D (2001) CMM measurement strategies. Natl Physical Laboratory, Teddingoton, Middlesex, United Kingdom
59. Hermann G (2008) Advanced techniques in the programming of coordinate measuring machines. In: Proceedings of the 6th international symposium on applied machine intelligence and informatics, IEEE, Herlany, pp 327–330
60. Legge DI (2001) Off-line programming of coordinate measuring machine. Licentiate thesis, Division of manufacturing engineering, Lulea University of Technology, Lulea
61. Lee JW, Kim MK, Kim K (1994) Optimal probe path generation and new guide point selection methods. Eng Appl Artif Intell 7(4):439–445
62. Gu P, Chan K (1996) Generative inspection process and probe path planning for coordinate measuring machines. J Manuf Syst 15(4):240–255
63. Lim PC, Menq HC (1994) CMM feature accessibility path generation. Int J Prod Res 32(3):597–618
64. Prieto F, Redarce T, Lepage R, Boulanger P (2002) An automated inspection system. Int J Adv Manuf Technol 19:917–925

65. Juster PN, Hsu HL, Pennington DA (1994) Advances in feature based manufacturing: the selection of surface for inspection planning. Elsevier
66. Lin YJ, Murugappan P (1999) A new algorithm for determining a collision free path for a CMM probe. Int J Mach Tools Manuf 39:1397–1408
67. Lin C-Z, Lin S-W (2001) Measurement point prediction of flatness geometric tolerance by using grey theory. Precis Eng J Int Soc Precis Eng Nanotechnol 25:171–184
68. Kweon S, Medeiros DJ (1998) Part orientations for CMM inspection using dimensioned visibility maps. Comput Aided Des 30:741–749
69. Ziemian CW, Medeiros DJ (1997) Automated feature accessibility for inspection on a coordinate measuring machine. Int J Prod Res 35(10):2839–2856
70. Lin JY, Murugappan P (2000) A new algorithm for CAD-directed CMM Dimensional inspection. Int J Adv Manuf Technol 16:107–112
71. Limaiem A, ElMaraghy AH (1997) Automatic planning for coordinate measuring machines. In: Proceedings of the 1997 IEEE, international symposium on assembly and task planning, Marina del Rey, CA, pp 243–248
72. Wu Y, Liu S, Zhang G (2004) Improvement of coordinate measuring machine probing accessibility. Precis Eng 28:89–94
73. Spitz NS, Spyridi JA, Requicha GAA (1999) Accessibility analysis for planning of dimensional inspection with coordinate measuring machines. IEEE Trans Robot Autom 15 (4):714–722
74. Alvarez JB, Fernandez P, Rico CJ, Mateos S, Suarez MC (2008) Accessibility analysis for automatic inspection in CMMs by using bounding volume hierarchies. Int J Prod Res 46 (20):5797–5826
75. Rico CJ, Valino G, Mateous S, Cuesta F, Suarez CM (2002) Accessibility analysis for star probes in automatic inspection of rotational parts. Int J Prod Res 40(6):1493–1523
76. Chiang YM, Chen FL (1999) CMM probing accessibility in a single slot. Int J Adv Manuf Technol 15:261–267
77. Weckenmann A, Estler T, Peggs G, McMurtry D (2004) Probing systems in dimensional metrology. Ann CIRP 53(2):657–684
78. Wozniak A, Dobosz M (2003) Metrological feasibilities of CMM touch trigger probes. Part I: 3D theoretical model of probe pretravel. Measurement 34(4):273–286
79. Jackman J, Park K-D (1998) Probe orientation for coordinate measuring machine systems using design models. Robot Comput-Integr Manuf 14:229–236
80. Moroni G, Polini W, Semeraro Q (1998) Knowledge based method for touch probe configuration in an automated inspection system. J Mater Process Technol 76:153–160
81. Ziemian W, Medeiros JD (1998) Automating probe selection and part setup planning for inspection on a coordinate measuring machine. Int J Comput Integr Manuf 11(5):448–460
82. Zhao F, Xu X, Xie SQ (2009) Computer-aided inspection planning—the state of the art. Comput Ind 60(7):453–466
83. Myeong WC, Honghee L, Gil SY, Jinhwa C (2005) A feature—based inspection planning system for coordinate measuring machines. Int J Adv Manuf Technol 26:1078–1087
84. Kramer RT, Huang H, Messina E, Proctor MF, Scott H (2001) A feature—based inspection and machining system. Comput-Aided Des 33(9):653–669
85. Zhang SG, Ajmal A, Wootton J, Chisholm A (2000) A feature based inspection process planning system for co-ordinate measuring machine (CMM). J Mater Process Technol 107:111–118
86. Takamasu K, Furutani R, Ozono S (1999) Basic concept of feature-based metrology. Measurement 26:151–156
87. Mohib A, Azab A, ElMaraghy H (2009) Feature-based hybrid inspection planning: a mathematical programming approach. Int J Comput Integr Manuf 22(1):13–29
88. Sundararajan V, Wright KP (2002) Feature based macroplanning including fixturing. J Comput Inf Sci Eng Trans ASME 2:179–191

89. Kamrani A, Nasr AE, Al-Ahmari A, Abdulhameed O, Mian HS (2014) Feature-based design approach for integrated CAD and computer aided inspection planning. Int J Adv Manuf Technol 76:2159–2183

90. Ramesh M, Hoi-Yip D, Dutta D (2001) Feature based shape similarity measurement for retrieval of mechanical parts. J Comput Inf Sci Eng Trans ASME 1:245–256

91. Shah JJ, Anderson D, Kim SY, Joshi S (2001) A Discourse on geometric feature recognition from CAD models. J Comput Inf Sci Eng Trans ASME 1:41–51

92. Hennann G (1997) Feature-based off-line programming of coordinate measuring machines. In: Proceedings of the 1997 IEEE international conference on intelligent engineering systems, IEEE, Budapest, pp 545–548

93. Pham TD, Martin FK, Khoo PL (1991) A knowledge-base preprocessor for coordinate-measuring machines. Int J Prod Res 29(4):677–694

94. Babic B, Nesic N, Miljkovic Z (2008) A review of automated feature recognition with rule-based pattern recognition. Comput Ind 59:321–337

95. Kim S, Chang S (1996) The development of the off-line measurement planning system for inspection automation. Comput Ind Eng 30(3):531–542

96. Cho WM, Lee H, Yoon SG, Choi HJ (2004) A computer-aided inspection planning system for on-machining measurement—Part II: Local inspection planning. KSM Int J 18:1358–1367

97. Lee H, Cho M-W, Yoon G-S, Choi J-H (2004) A computer-aided inspection planning system for on-machine measurement—Part I: Global inspection planning. KSME Int J 8 (18):1349–1357

98. Wong FSY, Chuah KB, Venuvinod PK (2006) Inspection process planning: algorithmic inspection feature recognition, and inspection case representation for CBR. Robot Comput-Integr Manuf 22:56–68

99. Wong YSF, Chuah BK, Venuvinod KP (2005) Automated extraction of dimensional inspection features from part computer-aided design models. Int J Prod Res 43(12):2377–2396

100. Li Y, Gu P (2004) Free-form surface inspection techniques state of the art review. Comput Aided Des 36:1395–1417

101. Hesham AH, Youssef MA, Shoukry KM (2012) Automated inspection planning system for CMMs. In: Proceedings of the international conference on engineering and technology, IEEE, pp 1–6

102. http://www.aiag.org/scriptcontent/index.cfm. Accessed 15 June 2017

103. Xu X, Newman ST (2003) Making CNC machine tools more open, interoperable and intelligent—a review of the technologies. Comput Ind 57(2):141–152

104. Horst J, Kramer T, Stouffer K, Falco J, Huang HM, Proctor F, Wavering A Distributed testing of an equipment-level interface specification. National Institute of Standards, Technology (NIST), Gaithersburg, Maryland, USA

105. Ahmed F (2013) Interoperability of product and manufacturing information (PMI) using ontology. Master thesis, Korea Advanced Institute of Science, Technology, Daejeon, Korea

106. Rippey W (2005) AIAG demonstrates metrology interoperability: to save you time and money. In: The international dimensional workshop, AIAG, Nashville

107. Stojadinovic S, Majstorović V (2010) Metrology interoperability. Total Qual Manage Excellence 38(4):83–89

108. William R (2005) AIAG demonstrates metrology interoperability: to save you time and money. In: The international dimensional workshop, AIAG, Nashville

109. http://www.cam-i.org/standards.html. Accessed 20 July 2016

110. I++DML version 1.6, Dimensional measurement equipment interface, Tutorial

111. Humienny Z (2009) State of art in standardization in GPS area. CIRP J Manuf Sci Technol 2(1):1–7

112. ISO/FDIS 10303-203: Industrial automation systems and integration—product data representation and exchange—Part 203: Application protocols: configuration controlled 3D design (2007)
113. ISO/FDIS 10303-224: Industrial automation systems and integration—product data representation and exchange—Part 224: Application protocol: mechanical product definition for process planning using machining features (2006)
114. STEP tools, Inc. http://www.steptools.com. Sept 2018
115. Germani M, Mandorli F, Mengoni M, Raffaeli R (2010) CAD-based environment to bridge the gap between product design and tolerance control. Precis Eng 34:7–15

Chapter 2
Ontological Knowledge Base for Integrating Geometry and Tolerance of PMPs

Abstract When engineering information is once created and applied, it is often stored and forgotten. Current approaches for information retrieval are not effective enough in understanding the engineering content, because they are not developed to share, reuse and represent information of the engineering domain [1]. This chapter presents the current state of engineering ontology (EO) development and proposes a new method for its development at conceptual level in order to reuse and share knowledge in the domain of coordinate metrology (CM) and inspection planning in that domain. More specifically, the method defines the development of ontology for the construction of knowledge base, as one of the basic components for integration of geometry and tolerance of PMPs.

2.1 Introduction

Inspection on a CMM is based on a complex software support for different classes of metrological tasks (tolerances). The design of a uniform inspection plan for a measuring part, whose inspection is to be performed on a CMM, today is a complex issue due to:

- the type of manufacturing and metrological complexity of a measuring workpiece,
- intuition and knowledge of the CMM inspection planner or programmer and
- model and software for a CAI model, as a part for the integrated design (CAD) system, machining planning (CAPP/CAM) and inspection planning, which is most frequently encountered today as a PLM system.

The mentioned problem can be solved by the development of ontology and ontological knowledge base to generate the conceptual inspection plan for a measuring part, based on which the automatic inspection planning for a concrete CMM, i.e. measuring protocol, can be generated.

Ontology development incorporates all necessary information modelled by using ontological components like classes, individuals, properties, class hierarchies and

© Springer Nature Switzerland AG 2019
S. M. Stojadinović and V. D. Majstorović, *An Intelligent Inspection Planning System for Prismatic Parts on CMMs*, https://doi.org/10.1007/978-3-030-12807-4_2

property hierarchies, whereas the ontological knowledge base provides an answer to preset questions with the help of so-called reasoners using predominantly the rules for the feature decomposition.

2.2   Engineering Ontology Development for the Domain of Inspection Planning in Coordinate Metrology

The term ontology is known from philosophy where it is defined as a branch of metaphysics that studies the nature of being or the kind of things that exist [2, 3]. In engineering, the term ontology is primarily related to knowledge presentation and reuse of knowledge. Besides the need for presentation and reuse of knowledge of an area, there is a need for sharing knowledge between different users (companies, laboratories, etc.).

Researchers in the field of artificial intelligence and the knowledge presentation emphasise that the main purpose of EO is transfer and exchange of knowledge. On the other side, some authors connect ontology to knowledge base pointing out that it presents the basic logical structure around which the knowledge base will be built [2, 4]. However, one thing is certain—the ontology has found its place in the areas where semantics is the base for communication between people and systems [5]. Some of the reasons that stimulate the development of methodologies for the development of engineering ontologies are:

- Today's engineers rarely make an effort to find engineering contents outside the search via keywords, ignoring at the same time the reuse of knowledge, because the appropriate tools for research of engineering information are not sufficiently developed [6].
- In the industrial sector, design engineers spend 20–30% of time communicating and assuming information [7].

Recently proposed ontological development in engineering can be categorised according to its purpose. As stated by [8], there are three purposes:

1. high level of knowledge specification of the domain,
2. the system of interoperability,
3. the exchange of knowledge and its reuse.

2.2.1   Methodologies for EO Development—General Case

The first attempt to consolidate the experiences gained in the development of ontology is presented in [9], emphasising the criteria such as clarity, coherence and extendibility. The work [10] considers the development of the enterprise ontology for modelling production processes within the company, and three strategies are proposed for concepts identification in ontology:

- top-down,
- bottom-up and
- middle-out.

Having in mind the complexity of the problem that engineers are facing, it is necessary to combine mentioned strategies for classes identification and in that way facilitate the approach to the concepts located 'in the middle', being the least approachable section of the hierarchy of classes. Ontological development and evaluation method applied for Toronto Virtual Enterprise (TOVE) ontology development [11] are based on a set of questions, so-called competent questions for determining the scope of ontology and extracting the major ontology classes. The TOVE method was developed to build a model based on the first-order logic for representing the ontology. Similar methods are reported in [12].

The work [13] presents the method for ontology development from the very beginning called METHONTOLOGY. However, its evaluation, in the opinion of experts in this field, is still subjective.

Among recently proposed methods for ontology development, some are adopted in engineering like:

- the application of the formal concept of analyses [14] to form the ontology of the family parts using image analysis, obtained with a disposable use of a camera,
- design of ontology development process adapted to the specific needs of the manufacturing companies [15].

However, the applied adoption is not an explicit study of specific relationships between concepts, and therefore, the result is a list of independent taxonomies, not ontologies.

The concept of engineering ontology development is founded on creating engineering lexicon, as a basis for further development of engineering ontology, and according to [8], it includes six steps (Fig. 2.1):

- specification,
- conceptualisation,
- formalisation,
- population,
- evaluation and
- maintenance.

The proposed concept of methodology development can be accepted only after ontology development and its successful implementation. An example of such ontology is skeletal—methodology proposed on the basis of experience in the enterprise ontology development—enterprise ontology [10]. For the development of ontology, different criteria are proposed and some of them are presented in [16].

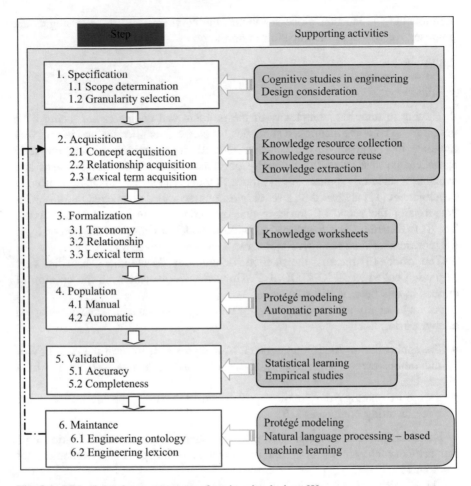

Fig. 2.1 EO and development process of engineering lexicon [8]

2.2.2 A Comparison of Traditional and New Approaches

Each development methodology of EO is specific per domain for which it develops. Since for the domain of coordinate metrology and metrology in general, almost no attempt has been made to develop EO, and no comparison can be drawn between traditional and new approaches. Relative comparison can be made from the viewpoint of general criteria such as clarity, coherence and extendibility for different domains and only after developing an ontology.

The ontological engineering is still in an early development stage and does not have detailed developed development methodology of EO and ontological characteristics that should be considered when ontology is developed. In summary, current methods for the development of ontologies require great efforts from those involved in their development, adoption and maintenance, aiming to integrate them.

Completeness and accuracy of the EO are reduced to evaluation of individual researchers or group of researchers who have developed EO, which is also understandable consequence of the specificity domain for which EO evolves.

2.2.3 Proposed Method of EO Development

To carry out the inspection of prismatic parts, among other things, we need data about their geometry and tolerance. Data on geometry contain CAD model of part in some of its output files such as IGES, STL or STEP. Perceived suitability of EO components to download necessary data from 3D CAD files and construction of the knowledge base creates a need for defining the method of EO development in the domain of coordinate metrology and inspection planning.

In the paper [12], the development process of an educational engineering ontology is defined in seven steps. Before we start developing one methodology, it is necessary to define the basic components of EO. According to [17], the basic components of ontology are:

- classes,
- individuals and
- properties.

Figure 2.2 shows the tagging principle of EO basic components for the domain of coordinate metrology.

The proposed method of EO development in this chapter is the result of a combined application of two concepts mentioned above and research carried out in [19]. The method consists of five steps, and its illustration is given in Fig. 2.3.

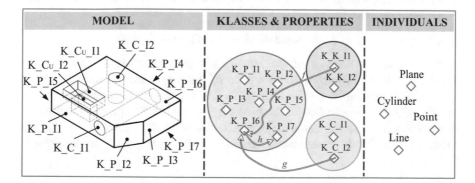

Fig. 2.2 A simpler example of geometry representation of a measuring workpiece via EO components: classes, properties and individuals [18]

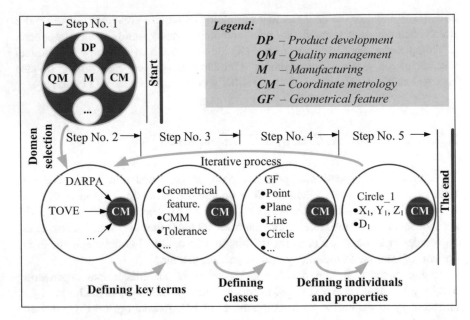

Fig. 2.3 One method of EO development for the domain of coordinate metrology and inspection planning for PMPs

Step 1. Determination of the field and scope of ontology includes defining:

- domain of ontology,
- purpose of ontology,
- maintenance of ontology.

One of the ways to determine the scope of the ontology is to create a list of the questions to which the ontological knowledge base should give answers [11]. The list of the questions and the answers to these questions help to improve the ontology in the early stages of development and to limit the scope of the information model of a certain domain.

Step 2. Consideration of the capabilities of existing ontologies.

It refers to the analysis of the possibilities for adaptation or acquisition of developed EO, mainly from the viewpoint of the scope and domain application. First of all, attention should be directed to defining the basic components of the existing EO; the organisation of the hierarchy of classes; its purpose: whether the domain is similar to the domain for which it is needed to develop the EO; etc. Libraries with already created ontologies for reusable Web ontologies are given in [20, 21].

Step 3. Enumeration of the important terms of the chosen domain.

In this stage of EO development, it is necessary to enumerate all possible terms that will be used in ontology development. Some of these terms will become names of classes, some will become names of class properties, and some will remain unused, because they are not important for defining the optimal scope of ontology. In this step, it is not considered whether a term belongs to some of the EO components and optimisation of the number of terms does not come to the fore, but the focus of attention should be not to leave out any of the terms.

Step 4. Defining of the classes and their hierarchy.

There are several possible approaches for the development of class hierarchy. In the paper [22], they are:

- Top–down: Development process begins with definition of the most general concept.
- Bottom–up: Development process starts from the most specific classes and their hierarchy.
- Combined: Development process that combines the previous two ways.

Illustration of all three approaches for defining the class hierarchy is shown in Fig. 2.4.

In the paper [17], classes are represented as the set of individuals. They can be organised in the superclass–subclass hierarchy, which is often called taxonomy (Fig. 2.5).

Step 5. Defining of individuals and properties.

It is said that one part of defined terms will be properties of the classes. According to [17], there are two main types of properties: properties of objects and properties of data. Properties of objects are relations between two individuals, and they can be:

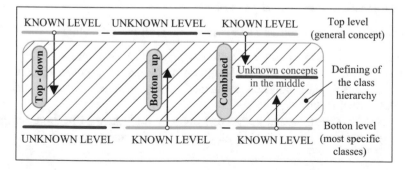

Fig. 2.4 Graphical representation of three approaches for creating the class hierarchy [18]

Fig. 2.5 Taxonomy [17]

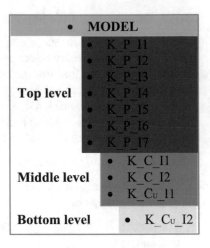

- inverse properties,
- functional properties,
- transitive properties,
- symmetric properties,
- antisymmetric properties,
- reflexive properties,
- irreflexive properties.

Detailed description of each of these properties is presented in the paper [17].

The last step in EO development implies defining individuals of all classes. Individuals represent objects in the field of interest. They are also referred to as instances and are the lowest possible level of representation in ontology. Defining individuals requires:

- selection of classes,
- creation of class instance,
- determination of individual properties.

Innovation of the proposed method is reflected in the following:

- decrease of interoperability between metrological software programmes,
- development of intelligent CMM segments: automatically generating the measuring sensor path, collision avoidance, precedence of inspection between metrological features,
- systematisation of domain knowledge done by a represented method, bridging the gap between CAD and CAI for the set of prismatic parts.

2.2.4 An Example of Method Implementation on the 3D CAD Model of Measuring Part

If it is assumed that the basic geometric primitives can be presented as EO classes, the description of one metrological part (Fig. 2.6) will be presented in this chapter from the viewpoint of the previously exposed method. The model of information about the ideal geometry covers the set of geometric information in relation to the measurement coordinate system of prismatic parts, which are presented as basic geometric features. The content of this set, using the presented method, is described by the basic components of EO, i.e. classes, individuals and properties. The basic approach of the method is defining the content of this set for inspection of prismatic parts on a CMM.

The set consists of:

- Classes: They represent the basic geometric primitives such as point (K_1), line (K_2), circle (K_3), ellipse (K_4), plane (K_5), sphere (K_6), cylinder (K_7), cone (K_8) and torus (K_9).
- Subclasses: Geometric features participating in the creation of other geometric features are subclasses of EO (K_11, K_12, K_13, ..., K_19; K_52, K_53, K_54, K_57, K_58, K_59).
- Individuals: They represent geometric features precisely defined by one or more parameters. An example of individuals for the point class is given with a label K_12_I1, K_13_I1, etc.
- Properties: Parameters of individuals represent the EO properties. There are four types of properties: coordinates, normal vectors, diameter and angle.

The proposed method includes all basic geometric features. Explicit application of the method implies data reusing, data sharing and logical structure of the knowledge base for an intelligent inspection of prismatic parts on a CMM. The main specificity of

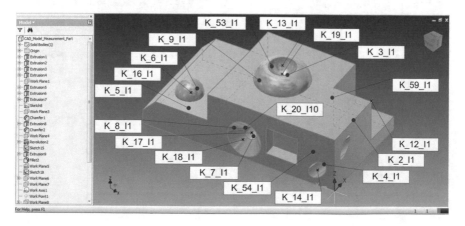

Fig. 2.6 Representation of metrological primitives as individuals of EO [18]

this approach is the possibility of describing each new prismatic part by using the already defined components of EO; however, the class hierarchy of a new part differs due to differences in the geometry and metrological complexity.

2.2.5 The Implementation in Software Protégé

Software Protégé is a free, open-source ontology editor and knowledge-based framework, relying on Java. Protégé implements a set of knowledge-modelling structures and actions that support the creation, visualisation and manipulation of ontology in various format representations [23].

The implementation in Protégé consists of modelling:

1. classes,
2. class hierarchy,

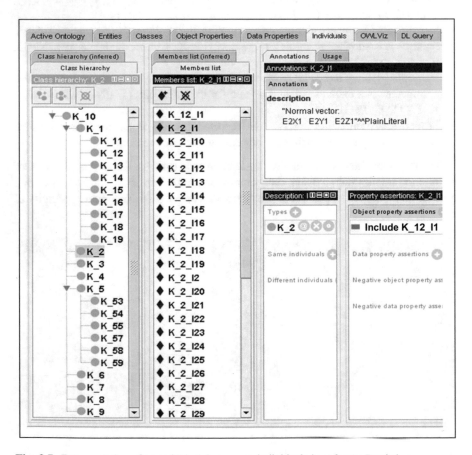

Fig. 2.7 Representation of metrological features as individuals in software Protégé

3. individuals,
4. class properties and,
5. individuals properties.

Classes represent geometric features like point, plane, circle, cylinder, cone, truncated cone, hemisphere and truncated hemisphere. Representation of the classes, subclasses, individuals and class properties in software Protégé is shown in Fig. 2.7. The set of classes is displayed in the window, on the left side. In Protégé, individuals are represented as higher specific classes, and in Fig. 2.7, they are shown in the middle window. For example, the point class consists of several individuals, which are points too but accurately defined by means of properties that in this case represent coordinates of the points in a specific coordinate system. As a result of the fact that the points take part in describing all geometric features, subclasses of points are defined, which contain individuals that describe ontologically all other primitives: K_1i, $= 2, 3, ..., 9$. Likewise, it is true for the plane, where subclasses K_53, K_54, K_55, K_57, K_58 and K_59 are defined.

2.3   Development of a Knowledge Base for the PMPs Inspection Planning on a CMM

Construction of the knowledge base model for inspection planning of PMPs on a CMM is a prerequisite for the development of intelligent inspection plan and its application in the concept of digital quality and digital manufacturing. Furthermore as it mentioned, digital manufacturing represents a framework for the development of a new generation of technological systems based on virtual simulation, product digital model and cloud computing concept application.

It started with the development of expert systems (ES) as a novel approach to modelling and applying engineering knowledge in technological systems, which had great expansion in the 1990s [24–27]. Since 2000 ESs have passed through the mature stage of development and application [27–32]. On the other hand, influenced by the development of ES for the technological processes planning (CAPP/CAM), especially STEP standardisation [33–35], the approach to metrological primitives modelling is based on the modelling of metrological characteristics (features) [36–38].

Investigations in this section are founded on the approach [35, 39, 40]. Overall, in a CAI model of ES, metrological features (metrological primitives) are considered in the following way: recognition and extraction from the CAD model, defining and modelling [38], and the approach emphasised herein has also evolved from defining and modeling [35], over ontology application for defining knowledge hierarchy for inspection [39] and integrated approach, which uses digital model of a product in interoperability environment (Autodesk Inventor 2011, PC-DMIS, Protégé).

2.3.1 Knowledge Base Model

The conceptual inspection plan presents a basis for CMM programming. Usually, based on experience, knowledge and skills, an engineer—inspection planner— generates a conceptual plan for inspection on CMMs. However, this approach should be avoided in a modern manufacturing, especially due to the fact that today CMMs work in a digital environment and with huge number of different parts. In contrast to this, a new concept is based on a usage of AI that generates conceptual inspection plan.

The first investigations in this domain were conducted a long time ago, defining a framework for the design of an inspection plan for CMMs using ES. Since then, owing to the development of artificial intelligence tools, a new generation of ESs has been developed. The base of each ES model in this domain is a knowledge base, its organisation, scope and content in terms of factual and heuristic knowledge [40]. A knowledge base must contain necessary knowledge and information about the measuring part (MP), its tolerances, geometric features (GF) and metrological features (MF), inspection sequences (IS), measuring probe configuration (PC), measuring machine (MM), and fixture tools (FT) and accessories, as presented in Fig. 2.8.

The basic steps for generating a conceptual inspection plan for CMMs according to [40] are:

- analysis and synthesis of metrological tasks (tolerances) according to the digital product model (the nodes MP, MF and GF presented in Fig. 2.8),
- definition and orientation of measuring coordinate systems (MP, MF, GF, IS, PC and MM),

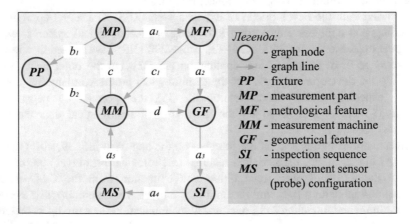

Fig. 2.8 Knowledge base graph [35, 40]

- selection of configuration of a measuring probe (IS, PC and MM) and
- definition of a measuring strategy (decomposition of metrological features, development of geometric features and inspection planning for them).

Therefore, our model is based on the following axiom: geometric (CAD)—technological (CAPP/CAM)—metrological (CAI) integration (Fig. 2.9), compatible with the state-of-the-art approaches in digital manufacturing [28].

2.3.2   An Experimental Example

An experiment is performed on a simple measuring part [40]. For the observed measuring part, a graph of knowledge base model is developed with four nodes and their interrelations, as presented in Fig. 2.10a. The nodes are non-terminated symbols presenting knowledge entities:

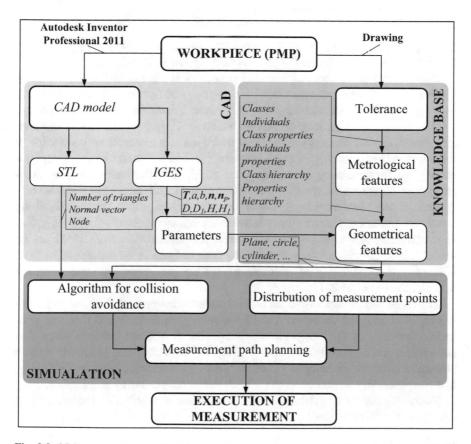

Fig. 2.9 Main elements for building the knowledge-based model for PMPs inspection on a CMM

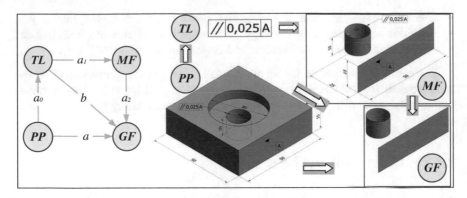

Fig. 2.10 Graph of a knowledge base and illustration on a simpler CAD model of the measuring part [40]

- MP—measuring part and its tolerances (TL),
- MF—metrological features,
- GF—geometric feature for the defined metrological feature.

Simple illustration of graph decomposition on the example of the parallelism tolerance of holes relative to the plane is also shown in Fig. 2.10b.

Non-terminated symbols as knowledge entities are connected with terminated symbols: a_0 = (tolerances, feature), a_1 = (metrological feature, feature, tolerances), a_2 = (geometric feature, features), b = (tolerance, geometric feature, features) and a = (measuring part, geometric features, features). They are presented as ontological structures with hierarchical relations that define all elements of knowledge in this domain. For example, relations between MP and GF could be a_0–a_1–a_2 (reasoning line for generating measuring probe path), a_0–b (reasoning line for generating point coordinates at the geometric feature) and a (reasoning line for geometric features at the measuring part). Some relevant investigations on the modelling based on tolerance characteristics have been already performed [33–35]. Our approach integrates metrological and geometric information from CAD digital model, where geometric information is parameters of geometric features taken from IGES file after modelling of prismatic measuring part using software Autodesk Inventor Professional 2011.

Therefore, the tolerances defined by ISO standard are presented by the graph that illustrates ontological decomposition into metrological features and then into geometric features. In this way, the reasoning line is defined, and, by extracting parameters, each geometric feature is uniquely defined. This geometric feature presents a basis for the analysis of accessibility of measuring probe, path planning, and for generating measuring protocols for the input measuring requirements.

Terminated symbol a_0 defines the decomposition of a prismatic measuring part into type of tolerances defined by ISO standard:

- length tolerances (TL),
- form/shape tolerances (TF),
- orientation tolerances (TO) and
- location tolerances (TLC).

Terminated symbol a_1 defines the continuation of decomposition into specific subtypes of tolerances defined in ISO standard. For example, the orientation tolerance could be decomposed into the following subtypes: parallelism (TO1), perpendicularity (TO2) and angularity (TO3).

Terminated symbol a_2 also defines the continuation of tolerances decomposition into the specific forms. These forms are denoted as metrological characteristics (features). From the metrological aspect, one metrological feature is composed of one or more geometric features. Figure 2.11 shows only one part of total ontological structures developed using software Protégé, which represents the rules for decomposition in a general case.

For example, the tolerance TLS is composed of six metrological features: TL1-1, TL1-2, TL1-3, TL1-4, TL1-5, TL1-6 and TL1-7. Further decomposition of each of them could be presented by five geometric features (GF100, GF200, GF500, GF600 and GF700).

At first, the geometric feature concept has been defined in analytical geometry. Based on this definition, it was later used in engineering modelling. In manufacturing engineering, the geometric features present a base for the definition of certain aspects of a product design, process design, inspection design, etc. In our model,

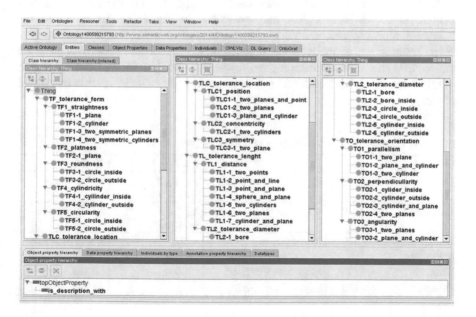

Fig. 2.11 Rules for decomposition of graph KB by Protégé [40]

the geometric feature presents the lowest level of tolerances definition or the object of generating the measuring probe points on the measuring part. The systematisation of geometric features is presented in next chapter.

The extraction of parameters of a geometric feature—cylinder—from IGES file is based on the recognition of its structure. A part of the structure needed for the analysis in our model is presented in Table 2.1.

As shown in Fig. 2.12, an IGES file is composed of five sections in the following order:

- start section,
- global section,
- directory entry section,
- parameter data section and
- terminate section.

The extraction of parameters (Table 2.1) is performed based on the number of sequences of an entity (geometric feature) and formulas for calculations given in Table 2.2. Using the same procedure as it was used for a cylinder, the extraction of parameters for the other geometric features is conducted.

The primary objective is a decomposition of a prismatic measuring part into metrological features (MF) that indirectly participate in the inspection planning. The secondary objective is a decomposition of tolerances into geometric features (GF), and, from metrological aspect, they give a full set of information to define a conceptual inspection plan.

An application of the before-mentioned concept of a knowledge base is performed on a real metrological part presented in Fig. 2.13. According to the aforementioned knowledge base model, tolerances of a part are reduced to geometric features, and all metrological features that are involved in part tolerances are enclosed. In that case, all metrological features taking part in the creation of part tolerance are involved. The forms of tolerances that a measuring part is composed of are:

Table 2.1 Extraction of IGES parameters of a cylinder [40]

Entity	1	1 2 3	5 6 7	73–80
Line (generatrix)	110	X_1, Y_1, Z_1 (start point)	X_2, Y_2, Z_2 (end point)	4 Seq. number
Line (axis)	110	X_3, Y_3, Z_3 (start point)	X_4, Y_4, Z_4 (end point)	5 Seq. number
Surface of revolution	120	Seq. no. 1, seq. no. 2	α_1, α_2 (start/end angle)	3 Seq. number
Direction	123	i_1, j_1, k_1 (unit vector)		18 Seq. number
Direction	123	i_2, j_2, k_2 (unit vector)		28 Seq. number

```
                                                                  S     1
,,8HCylinder,27HNo Iges File Name specified,22HAutodesk Inventor 2011,  G  1
37HAutoDesk Inventor Iges Exporter R2011,32,38,6,99,15,8HCylinder,1.0D0,G  2
2,2HMM,1,.08D0,15H20140414.165637,.01D0,10000.0D0,8HSLAVENKO,,11,0,,;  G  3
...
      110      4       0       0       0       0       0     000010000D  7
      110      0       0       1       0                           0D    8
      110      5       0       0       0       0       0     001010000D  9
      110      0       0       1       0                           0D   10
...
406,2,0,12HCylinder.ipt;                                            1P    1
143,0,5,2,11,17;                                                    3P    2
120,9,7,0.0D0,6.28318530717959D0;                                  5P    3
110,-7.5D0,0.0D0,47.25D0,-7.5D0,0.0D0,-2.25D0;                     7P    4
110,0.0D0,0.0D0,-2.25D0,0.0D0,0.0D0,-1.25D0;                       9P    5
...
123,0.0D0,0.0D0,1.0D0;                                             27P   18

123,-1.0D0,0.0D0,0.0D0;                                            45P   28
...
S      1G     3D      52P     32                                     T    1
```

Fig. 2.12 Part of the IGES file for a cylinder as a geometric primitive [40]

Table 2.2 Calculation parameters of a cylinder

Sequence number	Parameter of cylinder
4, 5	$D = ((X_6 - X_1)^2 + (Y_6 - Y_1)^2 + (Z_6 - Z_1)^2)^{0.5}$
5	$X_O = X_1,\ Y_O = Y_1,\ Z_O = Z_1$
5	$H = ((X_2 - X_1)^2 + (Y_2 - Y_1)^2 + (Z_2 - Z_1)^2)^{0.5}$
18	$n = [i_1\ j_1\ k_1]$

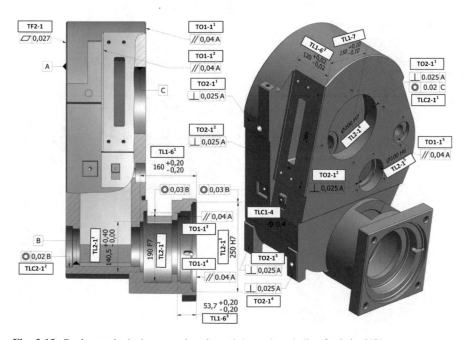

Fig. 2.13 Real metrological part—a housing of the main spindle of a lathe [40]

- tolerance of length (TL),
- tolerance of form (TF),
- tolerance of orientation (TO) and
- tolerance of location (TLC).

The length tolerance is composed of four length tolerances (TL1-61, TL1-62, TL1-62 and TL1-7) and five diameter tolerances (TL2-11, TL2-12 TL2-13 TL2-14 and TL2-15). The form tolerance is TF2-1, while the orientation tolerances are designated as TO2-11, TO2-12, TO2-13 TO2-14, TO2-11 and TO2-12. The location tolerance is marked as TLC1-4.

Figure 2.14 displays part of the ontological structure decomposition (class hierarchy, object property hierarchy and data property hierarchy) of a prismatic measurement part into geometric features. It started from the workpiece decomposition into general forms of tolerances defined by ISO standard. In this case, these tolerances are TL, TF, TO and TLC. These tolerance types are further decomposed

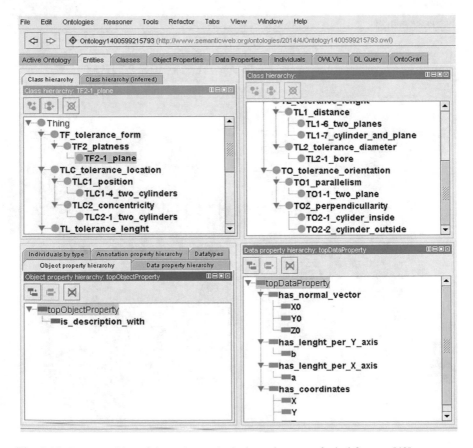

Fig. 2.14 Decomposition of the real metrological part into metrological features [40]

into specific types of tolerances that are also defined by the standard. It is necessary to follow this procedure in order to connect specific tolerance types with the real tolerances of real parts. The next iteration implies further decomposition into more specific forms of tolerances—forms found in technical drawing of a part.

Metrological features are composed of a few geometric features and present the relations between tolerance types and geometric features that PMP is composed of. In other words, if we know that the technical drawing of a part is a real source of metrological information, then the metrological features are introduced as a link between tolerances and a part geometry, whose carrier is a digital model of a part (CAD model). Software for CAD allows us to input a limited number of tolerances forms. This shows that, in inspection planning, CAD model of a part could be used only from the geometry aspect.

Inspection planning of prismatic parts on CMMs is usually performed with respect to three mutually orthogonal directions, depending on the number, position and orientation of measuring probes in a measuring sensor. This assumption serves as a base for the development of methods for defining a sequence of metrological features inspection for PMPs inspection on a CMM. From the mentioned three directions, it is possible in general to derive six directions of a measuring probe access (DPA). Since a measuring part must be set up at the machine table, one direction of access should be disregarded. Hence, there are five possible directions: DPA 1, DPA 2, DPA 3, DPA 4 and DPA 5 (Fig. 2.15). Each of these directions corresponds to some of the directions of a coordinate system of CMM: DPA 1 corresponds to direction $-Z$, DPA 2 corresponds to direction $-X$, DPA 3 corresponds to direction $-Y$, DPA 4 corresponds to direction $+X$, and DPA 5 corresponds to direction $+Y$.

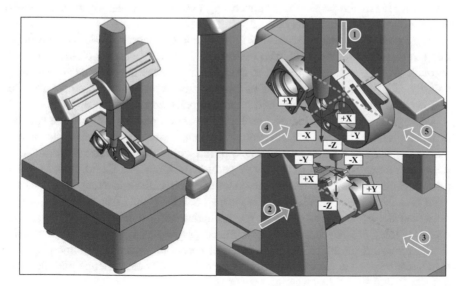

Fig. 2.15 Direction of probe accessibility for a three-axis CMM [40]

At the level of a geometric feature, DPA determines a vector **n** and a parameter **n**$_p$ which will be discussed in the next chapter. For the inspection of a feature, the possible directions are DPA 1, DPA 2, DPA 3, DPA 4 and DPA 5. A special case is when the direction of a vector **n** does not match any of the DPA directions. Consequently, the movement of a measuring probe should be decomposed into the movements in two DPAs. The first is the most approximate (matching criterion, the smallest angle) to the one of five above-mentioned DPAs for the movement directly in front of a geometric feature, and the second is vector direction of the **n** feature.

2.4 Concluding Remarks

On the basis of the analysis of the current state of methodology for engineering ontology development, a method of engineering ontology development at the conceptual level is proposed. Apart from reuse and knowledge distribution of a single domain, the developed method defines ontology development for the needs of creating knowledge base as one of the fundamental components of an intelligent system for PMPs inspection on a CMM. By defining engineering ontology with the help of the presented method, a set of terms is determined, which mapped into the domain of a knowledge base construction represents a set composed of basic components of the knowledge base, i.e. entities and relations between the entities.

The engineering ontology classes are the knowledge base entities, while relations between entities are engineering ontology properties. Explicit application of the method implies reuse, data distribution and logical structure of the knowledge base for intelligent inspection of prismatic parts on a CMM. Specificity of the approach reveals the possibility of describing each new prismatic part using engineering ontology components defined beforehand.

The result of the proposed method is an iterative process of ontology development in five steps for the coordinate metrology domain. Method implementation in software Protégé was performed on an example of a measuring part and shows great validity of the presented approach to engineering ontology development for the domain of coordinate metrology and prismatic parts inspection on a CMM.

Generation and application of a uniform inspection plan on a CMM are special problems which depend on metrological complexity of the prismatic parts, inspection planner's intuition and background knowledge. Research carried out provides a knowledge base model for inspection planning on a CMM so as to solve this problem and develop intelligent system for inspection planning. The knowledge base is defined by entities and relations between them. Also, the result of the presented approach is definition of the affiliation of geometric features to some forms of tolerances by browsing through the graph of a knowledge base model. When browsing through the graph, general forms of tolerances defined by the standard are linked to geometric features, so that it is possible to define metrological sequences and plan the measuring sensor path.

References

1. Li Z, Raskin V, Ramami K (2008) Developing engineering ontology for information retrieval. J Comput Inf Sci Eng 8:1–13
2. Swartout RW, Tate A (1999) Guest editors' introduction: ontologies. IEEE Intell Syst 14 (1):18–19
3. Chandrasekaran B, Josephson RJ, Benjamins RV (1999) What are ontologies, and why do we need them? IEEE Intell Syst 14(1):20–26
4. Martinez PS, Barreiro J, Cuesta E, Alvarez JB (2011) A new process based ontology for KBE system implementation: application to inspection process planning. Int J Adv Manuf Technol 57:325–339
5. Uschold M, Gruninger M (2004) Ontologies and semantics for seamless connectivity. SIGMOD Record 33(4):58–64
6. McMahon AC, Lowe A, Culley JS, Corderoy M, Crossland R, Shah T, Stewart D (2004) Waypoint: an integrated search and retrieval system for engineering documents. ASME J Comput Inf Sci Eng 4(4):329–338
7. Court WA, Ullman GD, Culley DGSJ (1998) A comparison between the provision of information to engineering designers in the UK and the US. Int J Inf Manag 18(6):409–425
8. Zhanjun L, Maria C, Karthik R (2009) A methodology for engineering ontology acquisition and validation. Artif Intell Eng Des Anal Manuf 23(1):37–51
9. Gruber T (1995) Towards principles for the design of ontologies used for knowledge sharing. Int J Hum Comput Stud 43(5–6):907–928
10. Uschold M, King M (1995) Towards a methodology for building ontologies. In: IJCAI95 workshop on basic ontological issues in knowledge sharing, Montreal
11. Gruninger M, Fox S (1995) Methodology for the design and evaluation of ontologies. In; Proceedings of international joint conference on ai workshop on basic ontological issues in knowledge sharing, Montreal
12. Noy NF, McGuinness DL (2001) Ontology development 101: a guide to creating your first ontology, knowledge systems laboratory and stanford medical informatics
13. Fernandez M, Sierra GAP (1999) Building a chemical ontology using METHONTOLOGY and the ontology design environment. IEEE Intell Syst 14(1):37–46
14. Nanda J, Simpson TW, Kumara SRT, Shooter SB (2006) A methodology for product family ontology development using formal concept analysis and web ontology language. ASME J Comput Inf Sci Eng 6(2):1–11
15. Ahmed S, Kim S, Wallace KM (2007) A methodology for creating ontologies for engineering design. ASME J Comput Inf Sci Eng 7(2):132–140
16. Kalfoglou Y (2001) Exploring ontologies. Handbook of software engineering and knowledge engineering, Singapore, vol 1, pp 863–887
17. Matthew H et al, A practical guide to building OWL ontologies using Protégé 4 and CO-DE tools. The University Of Manchester
18. Stojadinovic S, Majstorovic V (2014) Developing engineering ontology for domain coordinate metrology. FME Trans 42(3):249–255
19. Stojadinovic S, Majstorović V (2011) Metrological primitives in production metrology–ontological approach. In: Proceedings of the 34th international conference on production engineering, 29–30, Faculty of Mechanical Engineering Nis, Nis, Serbia, 28–30th Sept
20. http://www.ksl.stanford.edu/software/ontolingua/ (accessed 01.08.2016.)
21. http://www.daml.org/ontologies/ (accessed 01.03.2018)
22. Uschold M, Gruninger M (1996) Ontologies: principles, methods and applications. Knowl Eng Rev 11(2):1–69
23. http://protege.stanford.edu/ (accessed 01.07.2016)
24. ElMaraghy HA, Gu PH (1987) Expert system for inspection planning. Ann CIRP 36(1):85–89
25. Ziemian CW, Medeiros DJ (1997) Automated feature accessibility for inspection on a coordinate measuring machine. Int J Prod Res 35(10):2839–2856

26. Limaiem A, ElMaraghy AH (1997) Automatic planning for coordinate measuring machines. In: Proceedings of the 1997 IEEE, international symposium on assembly and task planning, 243–248, Marina del Rey, CA
27. Takamasu K, Furutani R, Ozono S (1999) Basic concept of feature-based metrology. Measurement 26:151–156
28. Stefano DP, Bianconi F, Angelo DL (2004) An approach for feature semantics recognition in geometric models. Comput Aided Des 36:993–1009
29. Moroni G, Polini W, Semeraro Q (1998) Knowledge based method for touch probe configuration in an automated inspection system. J Mater Process Technol 76:153–160
30. Mohib A, Azab A, ElMaraghy H (2009) Feature-based hybrid inspection planning: A mathematical programming approach. Int. J. Comput Integr Manuf 22(1):13–29
31. Wong FSY, Chuah KB, Venuvinod PK (2006) Inspection process planning: algorithmic inspection feature recognition, and inspection case representation for CBR. Robot Comput Integr Manuf 22:56–68
32. Wong YSF, Chuah BK, Venuvinod KP (2005) Automated extraction of dimensional inspection features from part computer-aided design models. Int J Prod Res 43(12): 2377–2396
33. Laguionie R, Rauch M, Hascoet JY, Suh SH (2011) An extended manufacturing Integrated System for feature-based manufacturing with STEP-NC. Int J Comput Integr Manuf 24 (9):785–799
34. Cho W-M, Seo I-T (2002) Inspection planning strategy for the on-machine measurement process based on CAD/CAM/CAI integration. Int J Adv Manuf Technol 19:607–617
35. Majstorovic DV (2003) Inspection planning on CMM based Expert System. In: Proceedings of the 36th CIRP international seminar on manufacturing system, 1–9, CIRP, Saarbrucken, Germany
36. Myeong WC, Honghee L, Gil SY, Jinhwa C (2005) A feature-based inspection planning system for coordinate measuring machines. Int J Adv Manuf Technol 26:1078–1087
37. Kramer RT, Huang H, Messina E, Proctor MF, Scott H (2001) A feature-based inspection and machining system, Comput Aided Des 33(9):653–669
38. Zhang SG, Ajmal A, Wootton J, Chisholm A (2000) A feature-based inspection process planning system for co-ordinate measuring machine (CMM). J Mater Process Technol 107:111–118
39. Stojadinovic S, Majstorović V (2012) Towards the development of feature-based ontology for inspection planning system on CMM. J Mach Eng 12(1):89–98
40. Majstorovic V, Stojadinovic S, Sibalija T (2015) Development of a knowledge base for the planning of prismatic parts inspection on CMM. Acta IMEKO 42(2):10–17

Chapter 3
The Model for Inspection Planning of PMPs on a CMM

Abstract This chapter presents a model of prismatic measurement parts (PMPs) inspection planning on CMMs, in terms of an intelligent system of inspection planning. The developed model is composed of mathematical model, inspection feature construction, sampling strategy, probe accessibility analysis, automated collision-free generation and probe path planning. The proposed model presents a novel approach for the automatic inspection and a basis for the development of an integrated, intelligent system of inspection planning. The advantages of this approach imply the reduction of preparation time due to an automatic generation of a measuring protocol, a possibility for the optimisation of measuring probe path, i.e. the reduction of time needed for the actual measurement and analysis of a workpiece setup, as well as an automatic configuration of measuring probes.

3.1 Introduction

In defining relationships between tolerances and geometry in previously mentioned two chapters, input information was specified for a model of PMPs inspection on a CMM based on basic geometric features. Apart from necessary information on features and their connection with tolerances, a single model for the offline inspection planning of PMPs contains some other elements which are a prerequisite for its development. Such elements include:

- measuring coordinate systems,
- configuration of the measuring probes,
- collision avoidance principle,
- module for metrological recognition of PMP,
- local inspection plan and
- global inspection plan.

As it is known, in performing measurements on a CMM the coordinate systems can be linked to a CMM itself, then to a measuring sensor, measuring part, as well

© Springer Nature Switzerland AG 2019
S. M. Stojadinović and V. D. Majstorović, *An Intelligent Inspection Planning System for Prismatic Parts on CMMs*, https://doi.org/10.1007/978-3-030-12807-4_3

as measuring task. Hence, there originates an established division of coordinate systems into:

* coordinate system of the CMM, coordinate system of the measuring sensor,
* coordinate system of the measuring part,
* coordinate system of the measuring task.

The role of coordinate systems in a model for the inspection of PMP does not differ from their primary role in other sciences (mathematics, physics, astronomy, etc.), and it implies connecting or describing sets of the coordinates. In this case, it is the matter of coordinates of the measuring points and other points which are elements of the measuring probe path, and they can be referred to as node points.

The collision avoidance principle in the context of the offline inspection planning for PMP generates a collision-free path of the measuring sensor in transition from one primitive to another and occurs as a consequence of automatic measuring path generation. Namely, in manual path planning the position of a probe and measuring workpiece is visually monitored, and thus, the probe is guided along the path without obstacles. Manual guidance of the probe requires inspection planner's intuition, whereas the planner's knowledge may influence the path length, to be shorter or longer.

Configuration of the measuring probes results from the analysis of the measuring sensor accessibility to a specific primitive or group of features with a tendency to minimise the number of changes of probes, and thereby to minimise the time for changes of probes.

The module for metrological recognition is based on the computer-supported design, CAD model of PMP in internal record (IGES file) and a feature modelling for inspection. Metrological recognition is actually reduction of geometric to metrological features, which has been done by means of ontological knowledge base in the previous chapter, whereas modelling of the features for inspection involves the definition of the parameters of features, so that each of them is uniquely defined. Modelled features represent a basis for a local inspection plan development. The local inspection plan defines automatic distribution of the measuring points for each feature separately, depending on the parameters it is uniquely defined by. Unlike local, the global inspection plan considers two or more parameters linking them by a model for the measuring sensor path planning. Hence, at the output, the initial path is obtained for a measuring sensor.

The inspection process of PMP based on an intelligent integrated model for the inspection planning of PMP on a CMM is given in Fig. 3.1.

Compared to the models developed to date, the main difference is in the achieved level of automation and autonomy of the inspection planning process. Human inevitable participation in the majority of decision-making operations decreases automation and autonomy level. Limitations in application are also among the criteria by which the models can be distinguished. Thus, there are models for rotating and prismatic parts, as well as for free-form surfaces. Majority of developed software programmes for inspection planning are not open for upgrading by

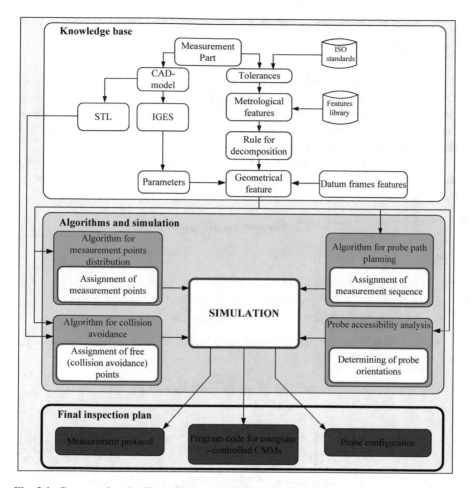

Fig. 3.1 Concept of an intelligent inspection of PMPs on CMM [1]

the user. This particularly refers to impossibility of the measuring path optimisation by applying some of the AI techniques (fuzzy logic, Hopfield neural networks, swarm theory, etc.). The path produced by the software is fixed for a chosen measuring strategy, and tolerances are loaded manually. The relationship between geometry (features) and tolerances for PMP does not exist in some form of record in the inspection planning process.

Main feature of the model is a system open for upgrading that will integrate all elements of the measuring process (measuring points distribution, probes config-uration, collision avoidance, etc.), with a minimum necessary human participation aiming to achieve a minimum measuring time of PMP on a CMM.

3.2 The Mathematical Model for Inspection Planning of PMPs on CMM

The component part of a model for the offline inspection planning of PMP on a CMM is a mathematical model shown in [1] (Fig. 3.2). It plays a major role in defining the relationships between coordinate systems of the measuring machine, measuring part and features. Each software producer for CMM possesses its own developed and implemented mathematical model.

Relying on the existing software for inspection on CMMs based on CAD representation of PMPs, the inspection principle is carried out based on the following formula:

$$^{M}\mathbf{r}_{P_i} = {}^{M}\mathbf{r}_W + {}^{W}\mathbf{r}_{P_i} \tag{3.1}$$

where

$^{M}\mathbf{r}_{P_i}$—probe point's position vector in machine coordinate system,
$^{M}\mathbf{r}_W$—workpiece system's position vector in machine coordinate system,
$^{W}\mathbf{r}_{P_i}$—probe point's position vector in workpiece coordinate system.

In the proposed model, the inspection principle for PMPs in MIP is based on the following equation:

$$^{M}\mathbf{r}_{P_i} = {}^{M}\mathbf{r}_W + {}^{W}\mathbf{r}_F + {}^{F}\mathbf{r}_{P_i} = {}^{M}\mathbf{r}_F + {}^{F}\mathbf{r}_{P_i} \tag{3.2}$$

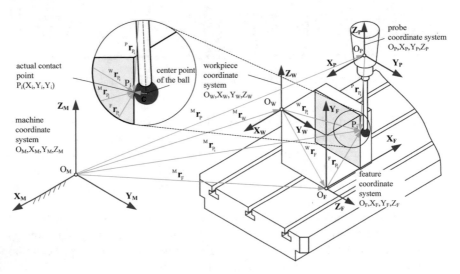

Fig. 3.2 Mathematical model inspection planning (MIP) for PMPs on CMM [1]

where

$^W\mathbf{r}_F$—feature system's position vector in workpiece coordinate system,
$^F\mathbf{r}_{P_i}$—probe point's position vector in feature coordinate system,
$^M\mathbf{r}_F$—feature (plane) system's position vector in machine coordinate system.

The remaining variables used in the calibration of measuring probes, i.e. in the establishment of relations between coordinate systems (CSs) O_P, X_P, Y_P, Z_P and O_M, X_M, Y_M, Z_M, are:

$^M\mathbf{r}_P$—vector of the probing system's reference point,
$^P\mathbf{r}_{P_i}$—probing system's position vector,
$\mathbf{c}$—tip correction vector (vector from the centre point to the actual contact point).

Observing Eqs. 3.1 and 3.2, the difference is noticeable implying the decomposition of vector $^W\mathbf{r}_{P_i}$ into the vector sum $^W\mathbf{r}_F + {}^F\mathbf{r}_{P_i}$. The observed difference is the result of incorporating primitive CS, i.e. the fact that the model of inspection is based on primitives to facilitate metrological recognition grounded on the CAD model of PMD. Coordinates of the measuring points $P_i(x_i, y_i, z_i)$, $i = 0, 1, 2, \ldots, (N-1)$ where N is the desired number of measuring points, are determined with regard to the coordinate system of a feature O_F, X_F, Y_F, Z_F, by the vector:

$$^F\mathbf{r}_{P_i} = x_{FP_i}\, \vec{i} + y_{FP_i}\, \vec{j} + z_{FP_i}\, \vec{k} \tag{3.3}$$

These coordinates are known in advance and generated based on the feature parameters taken from IGES file and sampling strategy (SS).

For the given vectors of measuring point positions $^F\mathbf{r}_{P_i}$, it is necessary to determine $^M\mathbf{r}_{P_i}$—probe point's position vector in machine coordinate system, and then, based on the Eq. 3.2, to determine coordinates of the measuring points with regard to the coordinate system of CMM. The vector $^M\mathbf{r}_{P_i}$ is determined using the matrix ^M_FT.

The position and orientation of a feature with regard to the coordinate system of a workpiece could be defined by the following matrix:

$$^W_FT = \begin{bmatrix} & ^W_FR & & \vdots & ^W\mathbf{r}_F \\ \hline 0 & 0 & 0 & \vdots & 1 \end{bmatrix} = \begin{bmatrix} i_{Fx} & j_{Fx} & k_{Fx} & \vdots & x_W \\ i_{Fy} & j_{Fy} & k_{Fy} & \vdots & y_W \\ i_{Fz} & j_{Fz} & k_{Fz} & \vdots & z_W \\ \hline 0 & 0 & 0 & \vdots & 1 \end{bmatrix} \tag{3.4}$$

The position and orientation of a workpiece with regard to the coordinate system of CMM are determined by the following matrix:

$$
{}^M_W T = \left[\begin{array}{ccc|c} & & & \\ & {}^M_W R & & {}^M\mathbf{r}_W \\ & & & \\ \hline 0 & 0 & 0 & 1 \end{array} \right] = \left[\begin{array}{ccc|c} i_{Wx} & j_{Wx} & k_{Wx} & x_M \\ i_{Wy} & j_{Wy} & k_{Wy} & y_M \\ i_{Wz} & j_{Wz} & k_{Wz} & z_M \\ \hline 0 & 0 & 0 & 1 \end{array} \right]
\tag{3.5}
$$

By multiplying the above two matrices expressed in relations (3.4) and (3.5), the matrix ${}^M_F T$ is obtained:

$$
{}^M_F T = {}^M_W T \cdot {}^W_F T = \left[\begin{array}{ccc|c} & & & \\ & {}^M_F R & & {}^M\mathbf{r}_F \\ & & & \\ \hline 0 & 0 & 0 & 1 \end{array} \right] = \left[\begin{array}{ccc|c} i_{MFx} & j_{MFx} & k_{MFx} & x_{MF} \\ i_{MFy} & j_{MFy} & k_{MFy} & y_{MF} \\ i_{MFz} & j_{MFz} & k_{MFz} & z_{MF} \\ \hline 0 & 0 & 0 & 1 \end{array} \right]
\tag{3.6}
$$

The next relation follows from (3.6):

$$
{}^M\mathbf{r}_F = x_{MF} \overrightarrow{i} + y_{MF} \overrightarrow{j} + z_{MF} \overrightarrow{k}
\tag{3.7}
$$

By replacing (3.3) and (3.7) in (3.2), we can obtain the following relation:

$$
{}^M\mathbf{r}_{P_i} = (x_{MF} + x_{FP_i}) \overrightarrow{i} + (y_{MF} + y_{FP_i}) \overrightarrow{j} + (z_{MF} + z_{FP_i}) \overrightarrow{k}
\tag{3.8}
$$

This procedure defines the probe point's position vector in the machine coordinate system for the ith measuring point, for the known vectors: ${}^M\mathbf{r}_W, {}^W\mathbf{r}_F$ and ${}^F\mathbf{r}_{P_i}$, $i = 0, 1, 2, \ldots, (N-1)$. The vector defines the selection of a coordinate system of a workpiece, and the vector ${}^W\mathbf{r}_F$ defines the position of a workpiece at the CMM granite table.

3.2.1   Inspection Feature Construction

A model for inspection feature construction (IFC) is based on the basic geometric features and their parameters, as presented in Fig. 3.3. The parameters of features presented in Fig. 3.3 are given in Table 3.1. Geometric primitives involved in this modelling are:

- point,
- plane,
- circle,
- hemisphere,
- cylinder,
- cone,
- truncated cone and
- truncated hemisphere.

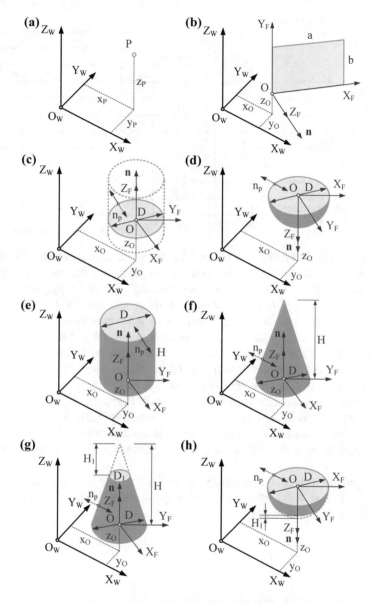

Fig. 3.3 Basic geometric features and their parameters: **a** point; **b** plane; **c** circle; **d** hemisphere; **e** cylinder; **f** cone; **g** truncated cone; **h** truncated hemisphere [1]

As above mentioned, the parameters of features define uniquely each feature. Defining parameters of the features was performed by fully describing their geometry, as well as whether the feature is full or empty. Defining of a full and empty feature is done based on the primitive fullness vector that will be explained in detail below, which provides information on whether the inspection of a given

Table 3.1 Parameters of the features

No.	Features	Parameters									
		${}^W_F T$	Feature CS								
			n	D	$\mathbf{n_p}$	H	D_1	H_1	a	b
1	Point	×								
2	Plane	×	×	×	×				×	×
3	Circle	×	×	×	×					
4	Hemisphere	×	×	×	×					
5	Cylinder	×	×	×	×	×				
6	Cone	×	×	×	×	×				
7	Truncated cone	×	×	×	×	×	×	×		
8	Truncated hemisphere	×	×	×	×		×	×		

primitive is performed inside or outside. The defined feature parameters are a basis for the development of algorithms such as the algorithm for measuring points' distribution, collision avoidance and path planning, where ontologically defined relationships between features and prescribed tolerances are also used.

Modelling of the features for inspection is also significant for optimisation by the ant colony application, as well as for defining the collision zones. The initial path obtained based on information (parameters) about features is the subject of optimisation by ant colony application in one of the chapters to follow.

The geometric feature term has been defined in analytical geometry and applied later in engineering modelling. In coordinate metrology, the set of geometric features presents a basis for defining IFC from the aspects of geometry and tolerances. In the inspection model based on IFC, geometric features present the lowest level of tolerance definition or the objects for generating the measuring sensor points at the workpiece.

Each geometric feature is uniquely determined by the local coordinate system O_F, X_F, Y_F, Z_F and a set of corresponding parameters. These parameters could belong to the following types:

- diameter (D, D_1),
- height (H, H_1),
- width (a),
- length (b),
- normal vector of a feature (**n**) and
- fullness vector of a feature ($\mathbf{n_p}$).

The vector **n** determines the orientation of a feature in the space. The fullness parameter is defined by a unit vector of the X-axis of a feature. The fullness vector $\mathbf{n_p} = \begin{bmatrix} -1 & 0 & 0 \end{bmatrix}$ defines a full feature, and a vector $\mathbf{n_p} = \begin{bmatrix} 1 & 0 & 0 \end{bmatrix}$ defines an empty feature. The fullness vector and the normal vector define the direction of a measuring probe access in generating the probe path.

The coordinate system O_F, X_F, Y_F, Z_F is determined with regard to the coordinate system O_W, X_W, Y_W, Z_W by parameters that are given as elements of the matrix $^W_F T$.

3.2.2   Sampling Strategy

An example of planning CMM sampling strategy based on a cost function is proposed in [2]. However, in cost function modelling the component of the planning of measurement strategy is not fixed. In general, this component depends on time required to perform the preparation of measurement and whether preparation is defined manually or automatically.

The approaches to defining the number of measuring points by means of fuzzy logic are given in [3, 4]. According to [3], the major criteria for defining the number of measuring points are the measuring surface of a feature, tolerance quality and machine accuracy.

Given the CMM accuracy and comparatively small dimensions of the measuring parts (measuring surfaces of features), the major criterion for us is tolerance quality. For desired number of measuring points and parameters of features at the input, based on the model of measuring points' distribution, to be described, the distributed measuring points are automatically obtained at the output. In that case, the desired number of points may imply any number, which also enables measurements of tolerances of shape such as circularity and cylindricity. Theoretical illustration of a method for points' distribution was performed for the number of points $N = 10$ for all features, while real measurements were realised for the numbers of points recommended in [5].

Sampling strategy (SS) is based on Hemmersly sequences presented in [6] for the calculation of coordinates for two axes of a feature:

$$s_i = \frac{i}{N}$$

$$t_i = \sum_{j=0}^{k-1} \left(\left[\frac{i}{2^j} \right] \text{Mod } 2 \right) \cdot 2^{-(j+1)}$$

where $k = \log_2 N$; N is the desired number of measuring points, $i = 0, 1, 2, \ldots, (N-1)$; Mod 2—mathematical operator whose result is the remainder of dividing by two. According to Hemmersly, the coordinate corresponding to X-axis is labelled with s_i, while along Y-axis with t_i.

The originally developed algorithm refers to 2D primitives, whereas afterwards, taking into account the third axis, the algorithm also considers 3D features such as the cone and hemisphere. However, the author for 3D features introduces indirectly the third coordinate term via 2D by the measuring points design and denotes them by w_i.

Since automatic distribution of measuring points is the priority, today this method has proved to be satisfactory in industry in terms of accuracy and scope of PMP tolerances. From the viewpoint of measuring inaccuracy, the author of the algorithm reports substantial improvement compared to distribution of uniform and random measuring points. The sequence is applicable for the distribution of measuring points only for basic geometric features but not for sculpture surfaces like the algorithm presented in [7].

By modifying the Hemmersly sequences, we define the distribution of measuring points for various geometric features that are involved in creation of PMP tolerances as follows:

To define the distribution of measuring points for a feature, the Cartesian coordinate system O_F, X_F, Y_F, Z_F and polar-cylindrical coordinate system O'_F, X'_F, Y'_F, Z'_F are needed. The coordinates in the Cartesian system are denoted by $P_i(s_i, t_i, w_i)$, and in polar-cylindrical coordinates system by $P_i(s'_i, t'_i, w'_i)$.

The equations for calculation of measuring point coordinates are:

- Plane:

$$s_i = \frac{i}{N} \cdot a$$

$$t_i = \left(\sum_{j=0}^{k-1} \left(\left[\frac{i}{2^j} \right] \mathrm{Mod}\, 2 \right) \cdot 2^{-(j+1)} \right) \cdot b$$

$$w_i = 0$$

where a (mm) is the plane limit for X-axis; b (mm) is the plane limit for Y-axis.

- Circle:

$$s_i = s'_i \cos\left(t'_i\right)$$
$$t_i = s'_i \sin\left(t'_i\right)$$
$$w_i = 0$$

where $s'_i = R$ and $t'_i = \frac{i}{N} \cdot 360°$.

- Hemisphere:

$$s_i = \sqrt{R^2 - \left(\left(\sum_{j=0}^{k-1} \left(\left[\frac{i}{2^j} \right] \mathrm{Mod}\, 2 \right) \cdot 2^{-(j+1)} \right) \cdot R \right)^2 \cdot \cos\left(\frac{i}{N} \cdot 360° \right)}$$

$$t_i = \sqrt{R^2 - \left(\left(\sum_{j=0}^{k-1}\left(\left[\frac{i}{2^j}\right]\mathrm{Mod}\,2\right)\cdot 2^{-(j+1)}\right)\cdot R\right)^2}\cdot \sin\left(\frac{i}{N}\cdot 360°\right)$$

$$w_i = \left(\sum_{j=0}^{k-1}\left(\left[\frac{i}{2^j}\right]\mathrm{Mod}\,2\right)\cdot 2^{-(j+1)}\right)\cdot R$$

where R (mm) is the radius of a hemisphere.

- Cylinder:

$$s_i = R\cos\left(-\frac{\pi}{2}-\frac{2\pi}{N}\cdot i\right)$$

$$t_i = R\sin\left(-\frac{\pi}{2}-\frac{2\pi}{N}\cdot i\right)$$

$$w_i = \left(\sum_{j=0}^{k-1}\left(\left[\frac{i}{2^j}\right]\mathrm{Mod}\,2\right)\cdot 2^{-(j+1)}\right)\cdot h$$

where h (mm) is the height of a cylinder.

- Cone, in polar-cylindrical coordinates:

$$s_i' = \left(1-\sum_{j=0}^{k-1}\left(\left[\frac{i}{2^j}\right]\mathrm{Mod}\,2\right)\cdot 2^{-(j+1)}\right)^{\frac{1}{2}}\cdot R$$

$$t_i' = \frac{i}{N}\cdot 360°$$

$$w_i' = (R-s_i')\cdot\frac{h}{R}$$

where R (mm) is the radius of the cone base, and h (mm) is the height of a cone.
In Cartesian coordinates, the formulations are $s_i = s_i'\cos(t_i')$; $t_i = s_i'\sin(t_i')$; $w_i = w_i'$.

- Truncated cone, in polar-cylindrical coordinates:

$$s_i' = R_1 + \left(1-\sum_{j=0}^{k-1}\left(\left[\frac{i}{2^j}\right]\mathrm{Mod}\,2\right)\cdot 2^{-(j+1)}\right)^{\frac{1}{2}}\cdot(R-R_1)$$

$$t_i' = \frac{i}{N} \cdot 360°$$

$$w_i' = (R - s_i') \cdot \frac{h_1}{R_1}$$

where R_1 (mm) is the radius of the smaller cone base, and h_1 (mm) is the height of a truncated cone.

In Cartesian coordinates, the formulations are $s_i = s_i' \cos(t_i')$; $t_i = s_i' \sin(t_i')$; $w_i = w_i'$.

- Truncated hemisphere:

$$w_i = \left(\sum_{j=0}^{k-1} \left(\left[\frac{i}{2^j} \right] \text{Mod}\, 2 \right) \cdot 2^{-(j+1)} \right) \cdot (R - h_1)$$

where h_1 (mm) is the height of a truncated hemisphere. Values s_i, t_i are calculated using the same formulas as for the hemisphere.

The distributions of measuring points for different features based on SS are presented in Fig. 3.4.

As above mentioned, the distribution was carried out for the number of measuring points $N = 10$ for basic geometric features such as point, plane, circle, cylinder, cone, truncated cone, truncated hemisphere. The presented formulas for distribution are subject to change in the number of measuring points, i.e. calculation can be performed for any N. Thus, the method of distribution offers many possibilities of inspection in terms of required PMP accuracy.

There are two major reasons for this selected strategy for distribution of measuring points:

- The first and major reason is the accuracy, that is, influence of points' distribution on measuring accuracy. Based on the results reported by Lee [6], the strategy selected in this reference, with respect to distribution of uniform and random measuring points, has a better accuracy. The conclusion was inferred by comparing root mean square error (RMSE) for various primitives.
- The second reason is the automatic aspect. Namely, for the desired number of measuring points and parameters of the features at the input, the distributed measuring points are automatically obtained at the output.

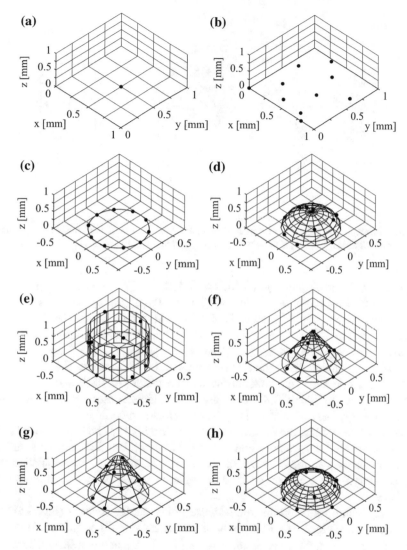

Fig. 3.4 Distribution of measuring points P_i for: **a** point; **b** plane; **c** circle; **d** hemisphere; **e** cylinder; **f** cone; **g** truncated cone; **h** truncated hemisphere

3.2.3 Probe Accessibility Analysis

The coordinates of measuring points are determined using sampling strategy (SS) as presented in the above section: $P_i(x_i, y_i, z_i), P_{i+1}(x_{i+1}, y_{i+1}, z_{i+1}), \ldots, P_{N-1}(x_{N-1}, y_{N-1}, z_{N-1})$ for $i = 0, 1, 2, \ldots, (N-1)$, with regard to O_F, X_F, Y_F, Z_F.

To perform an inspection, it is necessary to conduct the probe accessibility analysis (PAA). This analysis involves the determination of two sets of points:

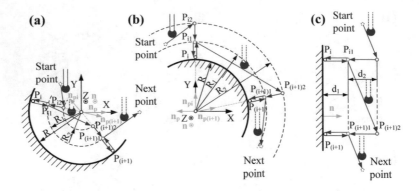

Fig. 3.5 Principle of definition of points P_{i1} and P_{i2} based on generated measuring points P_i: **a** concave cylindrical surface; **b** convex cylindrical surface; **c** flat surface [1]

$P_{i1}(x_{i1}, y_{i1}, z_{i1})$ and $P_{i2}(x_{i2}, y_{i2}, z_{i2})$ for $i = 0, 1, 2, \ldots, (N-1)$, as well as the definition of a fullness vector of a feature $\mathbf{n}_p$. The set $P_{i1}(x_{i1}, y_{i1}, z_{i1})$ presents points for the transition from rapid to slow feed of CMM. The distance between points $P_{i1}(x_{i1}, y_{i1}, z_{i1})$ and $P_i(x_i, y_i, z_i)$ is presented by d_1 that is a slow feed path, and the distance between points $P_{i2}(x_{i2}, y_{i2}, z_{i2})$ and $P_{i1}(x_{i1}, y_{i1}, z_{i1})$ is d_2—a rapid feed path. This approach (definition of the sets of points and feeds) enables the execution of PAA, in order to avoid the collision between a feature of PMP and a measuring probe. In the inspection of PMP, there are three different cases for the definition of point sets $P_{i1}(x_{i1}, y_{i1}, z_{i1})$ and $P_{i2}(x_{i2}, y_{i2}, z_{i2})$, as presented in Fig. 3.5.

As above mentioned, the fullness vector of a primitive has been introduced in the form of a vector for an empty and a solid primitive. The direction of a fullness vector coincides with the direction of X-axis of a feature, and the orientation with regard to the X-axis could be positive or negative, which defines the fullness of a primitive.

The procedure for determination of these point sets for concave (Fig. 3.5a) and convex surfaces (Fig. 3.5b) differs only in the orientation of a vector $\mathbf{n}_p$. In the first case—for concave surface the vector is $\mathbf{n}_p = \begin{bmatrix} 1 & 0 & 0 \end{bmatrix}$, and in a second case—for convex surface the vector is $\mathbf{n}_p = \begin{bmatrix} -1 & 0 & 0 \end{bmatrix}$. For convex and concave surfaces, the procedure for determination of the point sets $P_{i1}(x_{i1}, y_{i1}, z_{i1})$ and $P_{i2}(x_{i2}, y_{i2}, z_{i2})$ could be outlined into the following steps:

- **Step 1**: Formation of the vector $\overrightarrow{P_iO} = (x_0 - x_i)\vec{i} + (y_0 - y_i)\vec{j} + (z_0 - z_i)\vec{k}$, where the circle centre is determined by $O(x_0, y_0, z_0)$.

- **Step 2**: Formation of the vector $\mathbf{n}_{pi} = \dfrac{\overrightarrow{P_iO}}{\left|\overrightarrow{P_iO}\right|}$.

- **Step 3**: Formation of the vector $\overrightarrow{P_iP_{i1}} = \overrightarrow{n_{pi}} \cdot d_1 = x_{P_iP_{i1}}\vec{i} + y_{P_iP_{i1}}\vec{j} + z_{P_iP_{i1}}\vec{k}$ and $\overrightarrow{P_iP_{i2}} = \overrightarrow{n_{pi}} \cdot (d_2 + d_1) = x_{P_iP_{i2}}\vec{i} + y_{P_iP_{i2}}\vec{j} + z_{P_iP_{i2}}\vec{k}$, where for concave surface $d_1 = R - R_1$; $d_2 = R_1 - R_2$, and for convex surface $d_1 = R_1 - R$; $d_2 = R_2 - R_1$.

- **Step 4**: Calculation of the coordinates $P_{i1}(x_{i1}, y_{i1}, z_{i1})$:

$$
\begin{aligned}
x_{i1} &= x_{P_i P_{i1}} + x_i, \\
y_{i1} &= y_{P_i P_{i1}} + y_i, \\
z_{i1} &= z_{P_i P_{i1}} + z_i,
\end{aligned}
\tag{3.9}
$$

and coordinates point $P_{i2}(x_{i2}, y_{i2}, z_{i2})$ as

$$
\begin{aligned}
x_{i2} &= x_{P_i P_{i2}} + x_i, \\
y_{i2} &= y_{P_i P_{i2}} + y_i, \\
z_{i2} &= z_{P_i P_{i2}} + z_i.
\end{aligned}
\tag{3.10}
$$

For the flat surface (Fig. 3.5c), taking into account the fact that vector **n** has been taken over from the corresponding IGES file, the procedure for determination of $P_{i1}(x_{i1}, y_{i1}, z_{i1})$ and $P_{i2}(x_{i2}, y_{i2}, z_{i2})$ implies the formation of a vector $\overrightarrow{P_i P_{i1}} = \overrightarrow{n} \cdot d_1 = x_{P_i P_{i1}} \overrightarrow{i} + y_{P_i P_{i1}} \overrightarrow{j} + z_{P_i P_{i1}} \overrightarrow{k}$, $\overrightarrow{P_i P_{i2}} = \overrightarrow{n} \cdot d_2 = x_{P_i P_{i2}} \overrightarrow{i} + y_{P_i P_{i2}} \overrightarrow{j} + z_{P_i P_{i2}} \overrightarrow{k}$, where values for constants d_1 and d_2 are adopted depending on the measuring probe diameter. The coordinates are calculated using the formulas (3.9) and (3.10).

The coordinates of measuring points $P_{i1}(x_{i1}, y_{i1}, z_{i1})$ and $P_{i2}(x_{i2}, y_{i2}, z_{i2})$ based on the accessibility analysis are marked with blue and red colours and presented in Fig. 3.6. Distribution was carried out for the number of measuring points $N = 10$ for basic geometric primitives: point, plane, circle, hemisphere, cylinder, cone, truncated cone, truncated hemisphere.

The total path of a measuring probe, when N points are measured, is:

$$
D_{tot} = \sum_{i=0}^{N-1} \left(\left| \overrightarrow{P_{i2} P_{i1}} \right| + 2 \cdot \left| \overrightarrow{P_{i1} P_i} \right| + \left| \overrightarrow{P_{i1} P_{(i+1)2}} \right| \right)
\tag{3.11}
$$

where $\left| \overrightarrow{P_{i2} P_{i1}} \right|$ is rapid feed and $2 \cdot \left| \overrightarrow{P_{i1} P_i} \right|$ is double-crossed slow feed for the ith point, if $\left| \overrightarrow{P_{i1} P_{(i+1)2}} \right|$ is the path length in the probe transition from the preceded ith to the subsequent $(i + 1)$ node point. The vector form of this equation is:

$$
D_{tot} = \sum_{i=0}^{N-1} \left(\overrightarrow{P_{i2} P_{i1}} + 2 \cdot \overrightarrow{P_{i1} P_i} + \overrightarrow{P_{i1} P_{(i+1)2}} \right)
\tag{3.12}
$$

and represents the initial path of a measuring sensor in the inspection of PMP on a CMM for a single primitive. This path will be used afterwards as a basis for optimisation by the ant colony application.

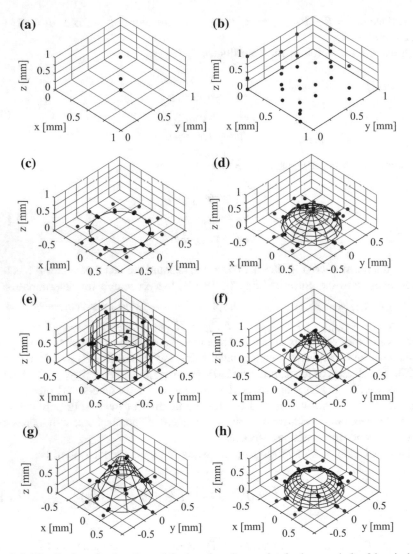

Fig. 3.6 Distribution of points P_{i1} and P_{i2} based on P_i: **a** point; **b** plane; **c** circle; **d** hemisphere; **e** cylinder; **f** cone; **g** truncated cone; **h** truncated hemisphere

3.3 Automated Collision-Free Generation

It is well known that 3D geometry can be represented in STereo Lithography (STL) format. In brief, this method of representing geometry consists of approximation of the nominal or CAD geometry of the measuring part with some number of triangles, each of them being described by coordinates of the three vertices. The surfaces of triangles represents an approximated area of the measuring part. In the context of path planning on a CMM, this format is used only for generating part of the measuring path, whose task is to avoid an obstacle in probe transition from one primitive to another; however, it is not used for describing primitives.

Based on STL model for the presentation of PMP geometry, the tolerances of PMP, the coordinates of the last point $P_{(N_{F1})1}$ of a feature F1 (precedent feature) and the coordinates of the first point $P_{(N_{F2})1}$ of a feature F2 (subsequent feature), the simplified principle of automated collision-free generation (ACFG) is presented in Fig. 3.7. The principle consists of iterative motion of the line p, which originally passes through points $P_{(N_{F1})1}$ and $P_{(N_{F2})1}$, for the value δ (mm) along Z-axis, according to thus defined procedure, as long as a line does not intersect the PMP volume. The line motion is done by translation for a given value. The problem of finding the intersection points between the line p and the PMP volume is reduced to the problem of finding an intersection point between the surface limited by a triangle and the line segment that connects points $P_{(N_{F1})1}$ and $P_{(N_{F2})1}$, and belongs to the line p.

For each triangle in STL file, the belonging plane equation is formulated. If triangle vertexes are T_1, T_2, T_3, the procedure of formation of the plane is described by the following equation:

$$Ax + By + Cz + D = 0 \qquad (3.13)$$

and it begins with the formation of a normal vector $\overrightarrow{n} = \overrightarrow{T_1 T_2} \times \overrightarrow{T_1 T_3} = A\,\overrightarrow{i} + B\,\overrightarrow{j} + C\,\overrightarrow{k}$, where from the constants A, B and C could

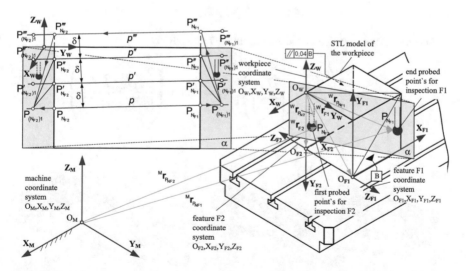

Fig. 3.7 Collision avoidance principle during the measuring probe crossover from a feature F1 to the next feature F2 [1]

be identified. The constant D is calculated using the scalar multiplication $D = -\vec{n} \cdot \vec{r_1}$, where $\vec{r_1} = \overrightarrow{OT_1}$. The next step is the formation of line equation through two points $P_{(N_{F1})1}$ and $P_{(N_{F2})1}$, based on the vector form of line equation:

$$\vec{M} = \vec{P} + t \cdot \vec{p} \tag{3.14}$$

where $\vec{p} = \overrightarrow{P_1 P_2}$, $\vec{P} = \overrightarrow{OP_1}$.

By projecting the Eq. (3.14) onto the axes of the Cartesian coordinate system, the line equation in a parametric form is obtained:

$$\begin{aligned} x &= x_0 + t \cdot p_x, \\ y &= y_0 + t \cdot p_y, \\ z &= z_0 + t \cdot p_z \end{aligned} \tag{3.15}$$

If an intersection between the line and the plane exists, then it is the point $P_j(x_j, y_j, z_j)$, where j is the number of intersection points. The coordinates of the mentioned point are obtained from the formulas (3.13) and (3.14).

Since $\Delta T_1 T_2 T_3$ is represented by a plane, and a line segment $\overline{P_{(N_{F1})1} P_{(N_{F2})1}}$ is represented as a part of a line p, it is necessary to check whether the intersection point P_j is placed at the line segment $\overline{P_{(N_{F1})1} P_{(N_{F2})1}}$ and whether it belongs to the part of the plane limited by $\Delta T_1 T_2 T_3$ (Fig. 3.8). This condition is encountered if a parameter $t \in (0, 1)$, and if the relation is satisfied:

$$\alpha \le \alpha_1 \wedge \beta \le \beta_1 \wedge \gamma \le \gamma_1.$$

The mentioned angles could be calculated as follows:

$$\alpha = \arccos\left(\frac{\overrightarrow{T_1 T_2} \cdot \overrightarrow{T_1 T_3}}{\left|\overrightarrow{T_1 T_2}\right| \cdot \left|\overrightarrow{T_1 T_3}\right|} \right), \quad \alpha_1 = \arccos\left(\frac{\overrightarrow{T_1 T_2} \cdot \overrightarrow{T_1 P_j}}{\left|\overrightarrow{T_1 T_2}\right| \cdot \left|\overrightarrow{T_1 P_j}\right|} \right)$$

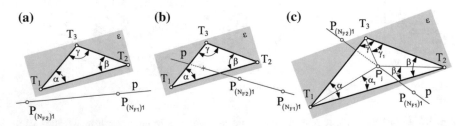

Fig. 3.8 Intersection between a line segment and a triangle: **a** a line segment does not intersect a triangle; **b** a line that forms the line segment intersects a triangle in a point that is not placed at the line segment; **c** a line segment intersects a triangle and intersection point is within a triangle [1]

$$\beta = \arccos\left(\frac{\overrightarrow{T_2T_1} \cdot \overrightarrow{T_2T_3}}{\left|\overrightarrow{T_2T_1}\right| \cdot \left|\overrightarrow{T_2T_3}\right|}\right), \quad \beta_1 = \arccos\left(\frac{\overrightarrow{T_2T_1} \cdot \overrightarrow{T_2P_j}}{\left|\overrightarrow{T_2T_1}\right| \cdot \left|\overrightarrow{T_2P_j}\right|}\right)$$

$$\gamma = \arccos\left(\frac{\overrightarrow{T_3T_2} \cdot \overrightarrow{T_3T_1}}{\left|\overrightarrow{T_3T_2}\right| \cdot \left|\overrightarrow{T_3T_1}\right|}\right), \quad \gamma_1 = \arccos\left(\frac{\overrightarrow{T_3T_2} \cdot \overrightarrow{T_3P_j}}{\left|\overrightarrow{T_3T_2}\right| \cdot \left|\overrightarrow{T_3P_j}\right|}\right)$$

If an intersection between the line p and any of the triangles from STL model exists, then using iterative procedure the points $P_{N_{F1}}, P_{N_{F2}}; P'_{N_{F1}}, P'_{N_{F2}};$ $P''_{N_{F1}}, P''_{N_{F2}}; \ldots; P^j_{N_{F1}}, P^j_{N_{F2}}$ are determined. The difference between the next point $P'_{N_{F1}}$ and the previous point $P_{N_{F1}}$ is in the value of z-axis, i.e. the value of the correction parameter δ (mm). The correction parameter is a constant for one PMP. The procedure is repeated until there are no remaining intersections between a line and triangles from STL model. The last points $P'''_{N_{F1}}$ and $P'''_{N_{F2}}$ of this iterative procedure are the points for which there is no collision during the measuring probe crossover.

By reduction of the number of triangles, the principle for collision avoidance results in a fast response, i.e. the coordinates of points without collision, and it does not significantly affect the processing time.

3.4 Probe Path Planning

Probe path planning (PPP) for the inspection of PMP on CMM depends on the geometric and metrological complexity of PMP. Geometric complexity refers to the arrangement and size of measuring surfaces, their accessibility for a measuring probe, etc. Metrological complexity refers to the shapes and quality of tolerances, as well as to their numbers. In coordinate metrology, i.e. CMM, these complexities must be considered together. The common element is the touching object, i.e. workpiece. Its position and orientation refer to geometric characteristics, and the tolerance of zone shapes, limits and referential elements refer to tolerance characteristics.

PPP presents a new way of inspection of PMPs, by defining the touching object to determine the tolerance and geometric aspects. Geometric information of the object is defined by IFC and taken over from IGES file. Its basis is a corresponding CAD file-modelled PMP. Integration of geometric and tolerance information is accomplished in the knowledge basis presented in [8]. This integration helps in defining the relations between the touching object and tolerances of PMP.

On the basis of defined integration of PMP tolerances and geometry and in previous chapters defined SS and the collision avoidance principle, PPP produces as an output the coordinates of points with their accurate sequence for path planning. Apart from measuring points, the node points through which the measuring probe passes in order to avoid collision are given. The output can also be an accurately defined sequence of a primitive inspection [9, 10].

The planning of an inspection of PMP on a CMM is performed with respect to three orthogonal directions. This fact is used for the definition of direction of a measuring probe access to PMP. The configuration of measuring probe is performed by comparing these directions with the directions of normal vectors. Therefore, apart from SS and the principle of collision avoidance, PPP contains the analysis of the accessibility of measuring probe for all PMP features.

References

1. Stojadinovic S, Majstorovic V, Durakbasa N, Sibalija T (2016) Towards an intelligent approach for CMM inspection planning of prismatic parts. Measurement 92:326–339
2. Moronia G, Petroa S (2013) Inspection strategies and multiple geometric tolerances. Procedia CIRP 10:54–60
3. Myeong CW, Honghee L, Gil YS, Jinhwa C (2005) A feature-based inspection planning system for coordinate measuring machines. Int J Adv Manuf Technol 26:1078–1087
4. Ha S, Hwang I, Rho HM (2000) A study on the development of knowledge based inspection planning system for CMM. In: 2nd CIRP international seminar on intelligent computation in manufacturing engineering, pp 497–502. CIRP, Capri, Italy
5. BS7172—Guide to assessment of position, size, and departure from nominal form of geometric feature, the minimum number of points
6. Lee G, Mou J, Shen Y (1997) Sampling strategy design for dimensional measurement of geometric features using coordinate measuring machine. Int J Mach Tools Manuf 37(7): 917–934
7. Chang HC, Lin CA (2009) An innovative algorithm for statistic sampling of measured points and simplifying measuring probe orientation for sculpture surfaces. Int J Adv Manuf Technol 41:780–798
8. Majstorovic VD, Stojadinovic SM (2013) Research and development of knowledge base for inspection planning prismatic parts on CMM. In: 11th international symposium on measurement and quality control, Cracow-Kielce, Poland, 11–13 Sept
9. Stojadinovic S, Majstorović V (2012) Determining the sequence of inspections of basic geometric primitives on CMM. In: Proceedings of the 38th JUPITER conference, Faculty of Mechanical Engineering Belgrade, Belgrade, 15–16 May, pp 5.25–5.30
10. Stojadinovic S (2016) Intelligent concept of inspection planning for prismatic parts on CMM. Doctoral dissertation (on Serbian language), University of Belgrade, Faculty of Mechanical Engineering, Belgrade

Chapter 4
The Model of Probe Configuration and Setup Planning for Inspection of PMPs Based on GA

Abstract This chapter presents an approach of probe configuration and setup planning for inspection of PMPs. The developed model is composed of two main parts: the analysis of PMP setups and the probe configuration for inspection on a CMM. A set of possible PMP setups and probe configurations for two types of sensors (probe star and probe head) is reduced to optimal number using a modified, current GA-based methodology. For each part setup, the optimal probe configuration and optimal point-to-point measuring path are possible to obtain. The advantage of the model is reduction of the total measurement time as well as elimination of errors due to human factor (minimising human involvement) through intelligent planning of probe configuration and part setup. This setup model can be applied not only for inspection planning on a CMM but also for the setup of prismatic parts machining on machining centres.

4.1 Introduction

In all former authors' investigations on the development of intelligent system for inspection planning of PMPs on a CMM [1–3], a detailed methodology was developed for obtaining optimal measuring path for a single, a priori known, part setup. In order to take into account all possible part setups and probe configurations, it is needed to perform the analysis of part setup and probe configuration, and thereafter by applying GA to reduce a set of possible solutions to optimal solution. The implementation of the analysis requires two data sets such as data on the geometry of PMPs and data on their tolerances.

To date, it has been a practice to take into account the measurement part geometry when generating a measuring protocol, or to enter the CMM programming software in the form of some output files (neutral data format, IGES or STEP, etc.), but to enter manually data on tolerances from a technical drawing. For now, there is no data format or file, where both data on geometry and data on tolerances are integrally stored. However, data format or application protocol that reveals the relationship between specified tolerances and part geometry involved in creating a

© Springer Nature Switzerland AG 2019
S. M. Stojadinović and V. D. Majstorović, *An Intelligent Inspection Planning System for Prismatic Parts on CMMs*, https://doi.org/10.1007/978-3-030-12807-4_4

given tolerance does not exist for now. For example, according to Zhao et al. [4], the current model of STEP data does not contain sufficient GD&T information for automatic generation of the inspection process planning. Clear definition of these data would facilitate intelligent inspection based on a CAD model of a part and eliminate use of a technical drawing as a basic medium of information on tolerances. Broadly viewed, this problem makes considerably hard the development of intelligent approach of probe configuration and setup planning for inspection of PMPs on a CMM, due to the division of a set of information required.

One of the solutions for the precursor of mentioned problems is viewed in the development of expert systems [5], and later in the development of an automatic inspection model with the GD&T elements [6–12].

One approach to defining the products and measurements can be feature-based and is presented in [13]. Each simple or compound feature may be one of the several types identifying several joint features in the measurement process planning—such as cones, cylinders and pattern-like features. An attempt to point to the importance of solution to these problems is presented in [14], and the development of an intelligent inspection planning system based on object-oriented programming is shown in [15]. The solution of mentioned problem is noticeable in the decomposition of a measurement part into geometric features and definition of their association with a certain form of tolerance. Furthermore, geometric features can be represented as classes of engineering ontology, as new techniques of artificial intelligence. In brief, the relationship between tolerances and geometric features can be represented via properties of the classes and individuals in a unique ontology database. The basic task of a database is to give metrological features at the output, that is, relationships between geometric features and tolerances for a measurement part. Integration of geometry and tolerances as a prerequisite for realisation of investigations in the present paper was carried out using a priori-developed ontology database of knowledge given in one of the previous chapters, as well as in [16].

Overall, mentioned problems create a gap between geometry and PMP tolerances and affect automatic generation of optimal number of part setups and optimal sensor configurations for the measurement process planning. For this reason, it is needed to develop intelligent approach of probe configuration and setup planning for inspection of prismatic measurement parts on a CMM, which is the subject of this chapter. The system is based on novel methodology of defining approach directions suitable for GA application and defining optimal number of probe setups and configurations based on a modified algorithm.

4.2 PMPs Setup Model

The analysis of PMPs setup and probe configuration is mainly integrated into accessibility analysis [17–24]. If viewed disintegrated, the basic difference is in that the analysis of setup is PMP-oriented, whereas the accessibility analysis is PMP-feature-oriented, with the needed probes tending to configure based on it. In

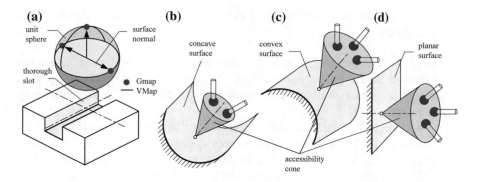

Fig. 4.1 Local feature accessibility [18, 22]: **a** sphere for thorough slot (GMap i VMap), **b** cone for concave surface, **c** cone for convex surface, **d** cone for planar surface

other words, for each PMP setup it is necessary to conduct accessibility analysis and based on it to configure the probes. Accessibility analysis is founded on approach directions that belong to the forms such as sphere (Fig. 4.1a) and cone (Fig. 4.1b–d).

For an ideal PMP, like cube or cuboid, the maximum number of setups equals the number of sides, i.e. six. Setup of parts with irregular shape was considered in [25].

On the one hand, the goal of analysis of the PMPs setup model in this subchapter is to satisfy technological measurement conditions (measurement bases, technological bases, dimensions of features, etc.), minimum number of measurement part setups, optimal probe configuration (minimum number of probes, etc.). On the other hand, the model must be adapted to an a priori-developed mathematical model (previous chapter) of inspection planning with the aim of integrating them and generating the point-to-point measuring path.

4.2.1   Mathematical Model of PMPs Setup

Mathematical model of PMPs setup should be built for defining a general matrix of the PMP setup, depending on the number of its features, and for applying GA technique to reduce a general matrix to the optimal set of PMP setups. The model is based on a priori-defined metrological and geometric features used to analyse setups and determine approach directions.

4.2.1.1   Geometric and Metrological Features in the Context of PMPs Setup

Geometric features and metrological features make up a basis for the analysis of the measurement sensor probes accessibility, measurement sensor configuring, path planning and generation of the measuring protocol for input measuring requirements.

As we have said, each geometric feature is unambiguously determined by a set of parameters relative to the local coordinate system (O_F, X_F, Y_F, Z_F) and measurement coordinate system of the measurement part (O_W, X_W, Y_W, Z_W). According to previous chapter and [1], the parameters can be of the following types: coordinates (X, Y, Z), diameter (D, D_1), height (H, H_1), width (a), length (b), feature vector (**n**), feature fullness vector (**n**$_p$).

Vector **n** determines feature orientation across space. Feature position is determined by coordinates (X_0, Y_0, Z_0). Fullness vector and feature vector define approach direction of the measurement sensor probe, whereas **n**$_G$ and **n**$_L$, respectively, represent the global approach direction for the feature and the local approach direction for the measuring point.

Vectors **n** and **n**$_p$, as well as newly introduced vectors **n**$_G$ and **n**$_L$, define a feature from the viewpoint of PMP setup and probe configuration. All mentioned vectors are shown in Fig. 4.2.

4.2.1.2   Approach Directions for PMPs

Approach directions for PMPs can be:

- Probe approach directions (PADs)—used as a basis for developing a probe configuration model and PMPs setup in the measuring process on a CMM.
- Feature approach direction (FAD)—defined as a potential approach direction for a feature from the standpoint of setup. It holds for FAD that it has the same direction as the PAD but opposite orientation.

Inspection planning for PMPs on a three-axis CMM, depending on the metrological complexity of PMPs, number, position and orientation of probes setup in a measurement sensor, is mainly performed from three mutually orthogonal directions corresponding to the Cartesian coordinate system. From mentioned three directions, it is possible to derive six potential PADs. Since it is needed to set up a PMP on a CMM table, one probe approach direction is omitted, so that five are left: PAD #1, PAD #2, PAD #3, PAD #4 and PAD #5 (Figs. 4.3 and 4.4).

Each of these directions has a direction corresponding to some of the CMM coordinate system directions, as follows:

- PAD #1—corresponds to direction $-Z$,
- PAD #2—corresponds to direction $-X$,
- PAD #3—corresponds to direction $-Y$,

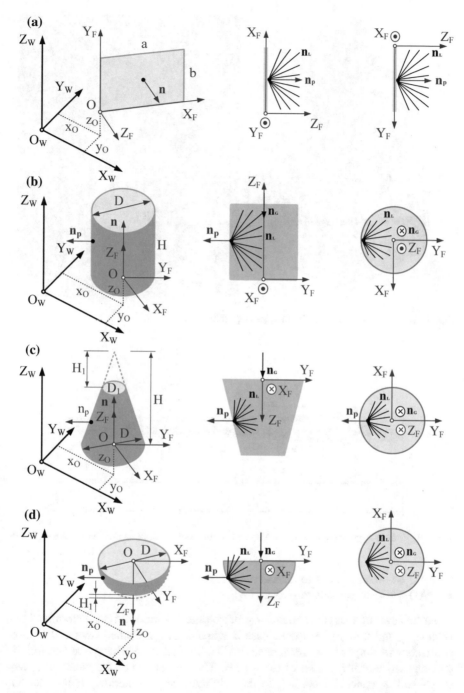

Fig. 4.2 Basic geometric features with parameters and defined local and global approach directions

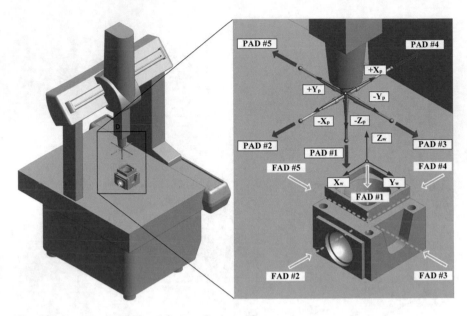

Fig. 4.3 An example PMP and features for inspection

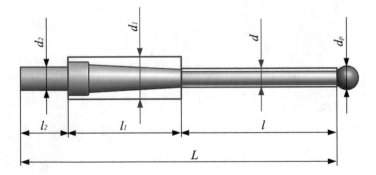

Fig. 4.4 Feature approach directions (FAD) and probe configuration approach directions (PAD)

- PAD #4—corresponds to direction $+X$ and
- PAD #5—corresponds to direction $+Y$.

At the level of a single geometric or metrological feature, PAD is determined by vectors $\mathbf{n_p}$ and $\mathbf{n}$ or $\mathbf{n_G}$. A special case is when direction of the vector $\mathbf{n}$ does not coincide with any of the mentioned PADs. In that case, the probe motion is decomposed into the motion of two PADs. The first is one of the mentioned five PADs and is selected according to the minimum angle criterion for the motion immediately in front of the geometric feature, while the second is in the direction of $\mathbf{n}$ feature.

Based on defined approach directions, a general form of matrices for the measurement part setup is defined as

$$S = \left[S_{ij}\right]_{m \times n} = S_{(i,j)} = S_{ij}((i,j) \in \{1, 2, \ldots, m\} \times \{1, 2, \ldots, n\})$$

respectively,

$$S = \begin{bmatrix} s_{11} & s_{12} & \cdots & s_{1n} \\ s_{21} & s_{22} & \cdots & s_{2n} \\ \vdots & \vdots & \ddots & \vdots \\ s_{m1} & s_{m2} & \cdots & s_{mn} \end{bmatrix} \tag{4.1}$$

where

m—number of features for inspection, or number of primitives that PMP is composed of,

$n = 6$—number of possible directions setup (approaches) for an ideal PMP model.

Matrix (4.1) includes all features, whose inspection is required, where $s_{ij} \in \{0, 1\}$ holds. So, matrix (4.1) has the form of Boolean matrix for whose elements it holds:

- $s_{ij} = 1$—feature can be approached from corresponding FAD direction,
- $s_{ij} = 0$—feature cannot be approached from corresponding FAD direction.

This manner of matrix definition for PMP setup encompasses a real fact that one feature can be approached from several FADs, which will be definitely used later in obtaining optimal probe configuration for selected PMP setup.

Matrix setup for test PMP is given by the expression (4.2)

$$S = \begin{bmatrix} 1 & 0 & 1 & 0 & 1 & 0 \\ 1 & 0 & 0 & 1 & 0 & 0 \\ 1 & 0 & 0 & 0 & 0 & 0 \\ 1 & 0 & 0 & 1 & 0 & 0 \\ 0 & 0 & 1 & 0 & 0 & 0 \\ 0 & 0 & 1 & 0 & 0 & 0 \\ 0 & 0 & 1 & 0 & 0 & 0 \\ 1 & 0 & 1 & 0 & 1 & 0 \\ 1 & 0 & 0 & 0 & 0 & 0 \\ 1 & 0 & 0 & 0 & 0 & 0 \\ 0 & 0 & 1 & 0 & 0 & 0 \\ 1 & 0 & 0 & 0 & 0 & 0 \\ 1 & 0 & 0 & 0 & 0 & 0 \\ 1 & 0 & 0 & 0 & 0 & 0 \\ 1 & 0 & 0 & 0 & 0 & 0 \\ 1 & 0 & 0 & 0 & 0 & 0 \end{bmatrix} \tag{4.2}$$

4.3   Probe Configuration Model for PMPs

The goal of the probe configuration model is optimal probe configuration for each part setup based on possible PADs. The configuration model links, on the one hand, CMM and its coordinate system and, on the other hand, PMP setup on a CMM granite table.

Since two types of finish measuring instruments are commonly used in inspection for PMPs on a CMM, the configuration model is divided into probe configuration of:

- probe star and
- probe head.

A general form of probe configuration matrices is defined as

$$C = \left[C_{ij} \right]_{m \times e} = C_{(i,j)} = C_{ij}((i,j) \in \{1, 2, \ldots, m\} \times \{1, 2, \ldots, e\})$$

respectively,

$$C = \begin{bmatrix} c_{11} & c_{12} & \cdots & c_{1e} \\ c_{21} & c_{22} & \cdots & c_{2e} \\ \vdots & \vdots & \ddots & \vdots \\ c_{m1} & c_{m2} & \cdots & c_{me} \end{bmatrix} \tag{4.3}$$

where

$e = (n - 1)$—total number of possible configurations.

Matrix (4.3) includes all possible probe configurations, where $c_{ij} \in \{0, 1\}$ holds. So, matrix (4.3) has the form of Boolean matrix for whose elements it holds:

- $c_{ij} = 1$—probe can perform inspection from corresponding PAD direction,
- $c_{ij} = 0$—probe cannot perform inspection from corresponding PAD direction.

The total number of possible configurations is smaller by one of the possible PADs, in a general case, due to PMP setup along that direction, and thereby the impossibility for a probe to perform inspection from one of the two directions.

4.3.1   Probe Star

In a general case, five probes with different orientations can be set up in the measurement sensor carrier. Defining the probe for the purpose of inspection planning process must fulfil two criteria:

- The first refers to the minimum sufficient number of parameters that define it in terms of volume in collision avoidance with PMP. A simplified model of one such probe is shown in Fig. 4.5.
- The second criterion should involve different probe orientations in the measurement sensor carrier, which requires decomposition of the star general case and coding of each case separately.

The thus obtained code represents one of the syllables of the GA population in the process of obtaining optimal configuration for a given PMP setup. A code

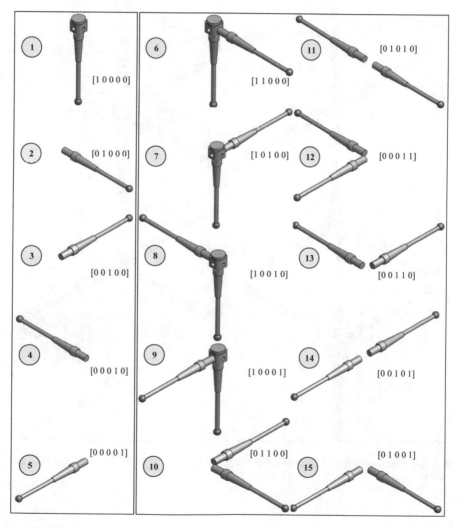

Fig. 4.5 Simplified probe model

possesses five digits $(e = (n - 1) = 5)$, which can be $c_{ij} \in \{0, 1\}$. The number of digits in a code equals the number of possible PADs.

Coding for the case of only one probe configuration and only two probes is demonstrated in Fig. 4.6, whereas coding for the case of configuring three, four and five probes is shown in Figs. 4.7 and 4.8.

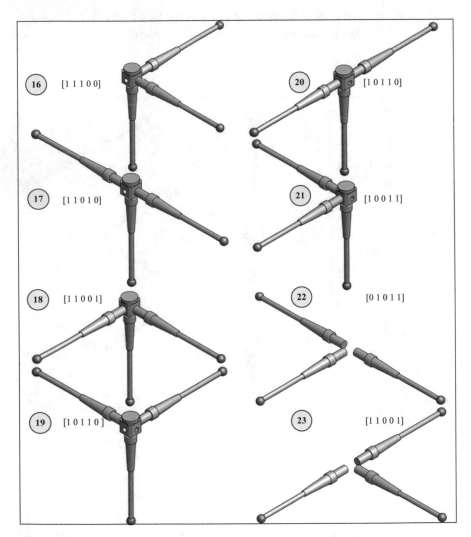

Fig. 4.6 Possible probe configuration directions for the case of using only one and only two probes

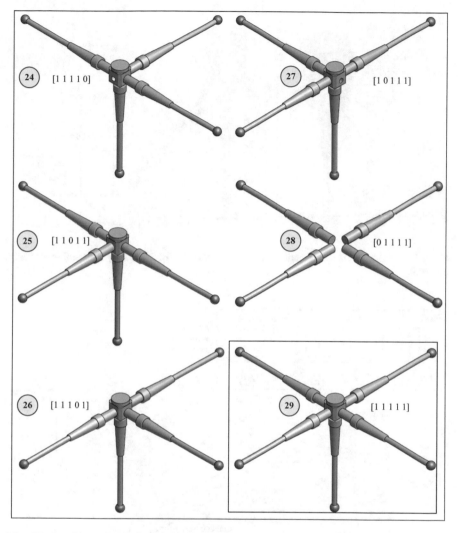

Fig. 4.7 Possible probe configuration directions for the case of using only three probes

4.3.2   *Probe Head*

As aforementioned, the second type of the measurement sensor can be a probe head. Its distinctive characteristic is 2DOFs in the form of two rotations (α and β). A probe head model is presented in Fig. 4.9a. The head is modelled like an industrial manipulator with 2DOFs (Fig. 4.9b). This modelling encompasses all possible probe head configurations, incrementally and continuously depending on its technical capacities. However, in order to arrive at the measurement sensor coordinate system, depending on the tip position (rubidium sphere) and probe

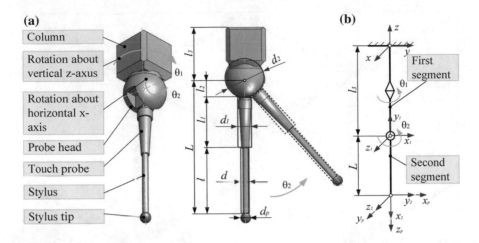

Fig. 4.8 Possible probe configuration directions for the case of using only four probes and only five probes

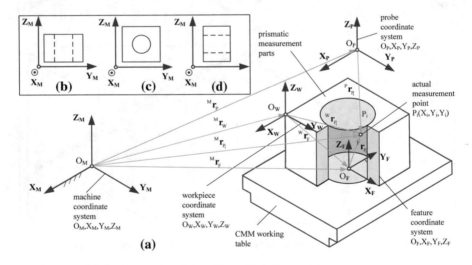

Fig. 4.9 Simplified probe head model: **a** 3D model, **b** line model

orientation of the head itself, a kinematic problem should be solved using Denavit–Hartenberg parameters (Table 4.1) and their algorithm of the manipulator segments coordinate systems adopted.

Table 4.1 Denavit—Hartenberg parameters of segments

Segment	α (°)	a (mm)	θ (°)	d (mm)
1	90	0	$\theta_1 + 90$	$-d_1$
2	0	a_2	$\theta_2 - 90$	0

Manipulator consists of two segments labelled 1 and 2. Segments of coordinate systems are, respectively, O_0, X_0, Y_0, Z_0—zero, O_1, X_1, Y_1, Z_1—coordinate system of first segment, O_2, X_2, Y_2, Z_2—coordinate system of second segment and O_p, X_p, Y_p, Z_p—coordinate system of the probe tip. Matrices determining segments position and orientation are ${}_1^0 A$ and ${}_2^1 A$.

$$
{}_1^0 A = \begin{bmatrix} {}_1^0 R & \vdots & {}^0 \mathbf{r}_1 \\ \cdots & \cdots & \cdots \\ 0 \quad 0 \quad 0 & \vdots & 1 \end{bmatrix} = \begin{bmatrix} c\theta_1 & -s\theta_1 \cdot c\alpha_1 & s\theta_1 \cdot s\alpha_1 & a_1 \cdot c\theta_1 \\ s\theta_1 & c\theta_1 \cdot c\alpha_1 & -c\theta_1 \cdot s\alpha_1 & a_1 \cdot s\theta_1 \\ 0 & s\alpha_1 & c\alpha_1 & d_1 \\ 0 & 0 & 0 & 1 \end{bmatrix} =
$$

$$
(4.4)
$$

$$
= \begin{bmatrix} -s\theta_1 & 0 & c\theta_1 & 0 \\ c\theta_1 & 0 & s\theta_1 & 0 \\ 0 & 1 & 0 & -d_1 \\ 0 & 0 & 0 & 1 \end{bmatrix}
$$

$$
{}_2^1 A = \begin{bmatrix} {}_2^1 R & \vdots & {}^1 \mathbf{r}_2 \\ \cdots & \cdots & \cdots \\ 0 \quad 0 \quad 0 & \vdots & 1 \end{bmatrix} = \begin{bmatrix} c\theta_2 & -s\theta_2 \cdot c\alpha_2 & s\theta_2 \cdot s\alpha_2 & a_2 \cdot c\theta_2 \\ s\theta_2 & c\theta_2 \cdot c\alpha_2 & -c\theta_2 \cdot s\alpha_2 & a_2 \cdot s\theta_2 \\ 0 & s\alpha_2 & c\alpha_2 & d_2 \\ 0 & 0 & 0 & 1 \end{bmatrix} =
$$

$$
(4.5)
$$

$$
= \begin{bmatrix} s\theta_2 & c\theta_2 & 0 & a_2 s\theta_2 \\ -c\theta_2 & s\theta_2 & 0 & -a_2 c\theta_2 \\ 0 & 0 & 1 & 0 \\ 0 & 0 & 0 & 1 \end{bmatrix}
$$

Position and orientation of second segment relative to zero segment are determined by the following matrix.

$$
{}_2^0 T = {}_1^0 A \cdot {}_2^1 A = \begin{bmatrix} -s\theta_1 \cdot s\theta_2 & -s\theta_1 \cdot c\theta_2 & c\theta_1 & -a_2 s\theta_1 s\theta_2 \\ c\theta_1 \cdot s\theta_2 & c\theta_1 \cdot c\theta_2 & s\theta_1 & a_2 c\theta_1 s\theta_2 \\ -c\theta_2 & s\theta_2 & 0 & -a_2 c\theta_2 - d_1 \\ 0 & 0 & 0 & 1 \end{bmatrix}
$$

$$
(4.6)
$$

The probe tip position and probe orientation relative to second segment are determined by the following matrix.

$$
{}^2_cT = \begin{bmatrix} {}^2_pR & \vdots & {}^2\mathbf{r}_c \\ \cdots\cdots & \vdots & \cdots \\ 0 \quad 0 \quad 0 & \vdots & 1 \end{bmatrix} = \begin{bmatrix} 0 & 0 & 1 & \vdots & 0 \\ 1 & 0 & 0 & \vdots & 0 \\ 0 & 1 & 0 & \vdots & 0 \\ \cdots & \cdots & \cdots & \vdots & \cdots \\ 0 & 0 & 0 & \vdots & 1 \end{bmatrix} \tag{4.7}
$$

The probe tip position and probe orientation relative to zero segment are determined by the following matrix.

$$
{}^0_2T = {}^0_2T \cdot {}^2_cT = \begin{bmatrix} -s\theta_1 \cdot c\theta_2 & c\theta_1 & -s\theta_1 s\theta_2 & \vdots & -a_2 s\theta_1 s\theta_2 \\ c\theta_1 \cdot c\theta_2 & s\theta_1 & c\theta_1 s\theta_2 & \vdots & a_2 c\theta_1 s\theta_2 \\ s\theta_2 & 0 & -c_2 & \vdots & -a_2 c\theta_2 - d_1 \\ \cdots & \cdots & \cdots & \vdots & \cdots \\ 0 & 0 & 0 & \vdots & 1 \end{bmatrix} \tag{4.8}
$$

After matrix (4.8) is obtained, a setup mathematical model and a probe configuration model can be integrated into a mathematical model of path planning as presented in the previous chapter. The aim of integration is to complete intelligent path planning system for PMP on a CMM in such a way so that both PMP setup and probe configuration are taken into account.

Integration is done using matrix ${}^M_FT$ from the previous chapter, which defines feature position and orientation relative to the measurement machine coordinate system (CS):

$$
{}^M_FT = {}^M_WT \cdot {}^W_FT = \begin{bmatrix} {}^M_FR & \vdots & {}^M\mathbf{r}_F \\ \cdots\cdots & \vdots & \cdots \\ 0 \quad 0 \quad 0 & \vdots & 1 \end{bmatrix} = \begin{bmatrix} i_{MFx} & j_{MFx} & k_{MFx} & \vdots & x_{MF} \\ i_{MFy} & j_{MFy} & k_{MFy} & \vdots & y_{MF} \\ i_{MFz} & j_{MFz} & k_{MFz} & \vdots & z_{MF} \\ 0 & 0 & 0 & \vdots & 1 \end{bmatrix} \tag{4.9}
$$

where ${}^M_WT$—matrix defining PMP position and orientation relative to the measurement machine CS and ${}^W_FT$—that defines feature position and orientation relative to the workpiece CS or PMP.

In a general case, matrix ${}^M_FT$ is variable for the set of possible setups and includes the entire set (number) of PMP setups. Furthermore, a set of possible setups makes matrix ${}^M_WT$ variable due to change of the workpiece CS position and orientation, and thereby the matrix ${}^M_FT$ too, due to change of the feature CS position and orientation. However, ${}^W_FT$ matrix remains unchanged, with the condition that there is no change of the workpiece CS. This can be explained based on Fig. 4.10, where a set of possible setups for a given PMP is presented.

The mathematical model developed in [1] takes into account a single setup (Fig. 4.10a) and describes a measuring path for it using a vector (4.10)

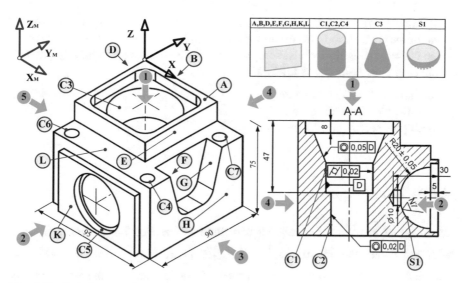

Fig. 4.10 PMP setup model

$$^M\mathbf{r}_{P_i} = \left(x_{MF} + x_{FP_i}\right)\overrightarrow{i} + \left(y_{MF} + y_{FP_i}\right)\overrightarrow{j} + \left(z_{MF} + z_{FP_i}\right)\overrightarrow{k} \qquad (4.10)$$

The developed setup and configuration model, in the present research, extend the given model to a general set of setups and configurations, which is given in Fig. 4.10b–d for this PMP, so that based on relation (4.10) the measuring path can be automatically defined for each new setup.

Using the algorithm to be presented in the next subchapter, a general set is reduced to optimal number of PMP setups and probe configurations. The extended mathematical model has created conditions for applying GA as an optimisation technique to obtain minimum number of PMP setups and probe configurations for each of them, but which together enable a complete PMP inspection. Afterwards, for each setup, initial measuring paths are defined by means of a model for measuring points distribution and collision avoidance presented in [1]. The thus obtained initial path is optimised by applying ACO method described in [2].

4.4 GA Model

The application of GA in the context of PMPs setup and probe configuration is a novelty, but in the context of path planning and its optimisation the results are reported in [26, 27].

The methodology described in the above two subchapters yield, at the output, Boolean matrices $\left[S_{ij}\right]_{m \times n}$ and $\left[C_{ij}\right]_{m \times e}$ or a set of possible problem solutions to

PMP setup and configuration. To arrive at the optimal solution, it is needed to apply some of the population-based optimisation methods. Considering that in this case it is the matter of discrete optimisation problem, GA is suitable for it. Simply put, GA consists of three sections, as follows:

1. definition of initial population,
2. definition of fitness function,
3. selection, crossover and mutation.

4.4.1 Initial Population

Based on the setup analysis, coding was done for each of the geometric features with six digits and matrix $\left[S_{ij}\right]_{m \times n}, ((i,j) \in \{1,2,\ldots,m\} \times \{1,2,\ldots,n\})$ was obtained. Then, using a probe configuration model, matrix $\left[C_{ij}\right]_{m \times e}, ((i,j) \in \{1,2,\ldots,m\} \times \{1,2,\ldots,e\})$ was obtained. If we introduce a new matrix P, which represents the matrix of randomly defined rows of matrix C, it can be defined as the GA initial population, and modified optimisation algorithm presented by Rice and Nyman [28] can be applied, which has Boolean matrix for the initial population and optimal solution. The difference in defining the initial population relative to the algorithm [28] is that here it is not randomly defined but, as above mentioned, it is obtained based on matrix C. The initial population matrix for test PMP is given by the following expression:

$$
P = C = \begin{bmatrix}
1 & 0 & 1 & 0 & 1 \\
1 & 0 & 0 & 1 & 0 \\
1 & 0 & 0 & 0 & 0 \\
1 & 0 & 0 & 1 & 0 \\
0 & 0 & 1 & 0 & 0 \\
0 & 0 & 1 & 0 & 0 \\
0 & 0 & 1 & 0 & 0 \\
1 & 0 & 1 & 0 & 1 \\
1 & 0 & 0 & 0 & 0 \\
1 & 0 & 0 & 0 & 0 \\
0 & 0 & 1 & 0 & 0 \\
1 & 0 & 0 & 0 & 0 \\
1 & 0 & 0 & 0 & 0 \\
1 & 0 & 0 & 0 & 0 \\
1 & 0 & 0 & 0 & 0 \\
1 & 0 & 0 & 0 & 0
\end{bmatrix} \tag{4.11}
$$

4.4.2 Fitness Function

Modification of the algorithm presented by Rice and Nyman [28] also refers to the change in fitness function and manner of gene selection, which provides for obtaining the optimal number of probe setups and configurations. A new fitness function in the matrix form is

$$F = \left[F_{1ij} \right]_{m \times n} \cdot \left[F_{2ij} \right]_{m \times n} \tag{4.12}$$

where

$$F_1 = \left[F_{1ij} \right]_{m \times n} = F_{(1i,j)} = \left[\sum_{i=1}^{m} c_{ij} \right], \quad j = 1, 2, \ldots, n$$

$$F_2 = \left[F_{2ij} \right]_{m \times n} = F_{(2ij)} = \left[\sum_{j=1}^{n} c_{ij} \right], \quad i = 1, 2, \ldots, m \tag{4.13}$$

The fitness function, being one of the major sections of the algorithm, should fulfil, in this case, two conditions. The first condition is to take from each row of the initial population P matrix at least 1 unit and at most 5 units each. The second condition is that this unit is taken from the column, whose sum is maximum. This way, priority columns or columns whose sum is maximum and less important columns or non-priority columns are extracted. To increase the difference and separate clearly one column from another column, the product of matrices is used. The principle of defining the fitness function also has physical meaning that can read like this: possible approach directions (probe configuration) for a single PMP setup become ultimate (optimal) only along the most represented directions.

4.4.3 Selection, Crossover and Mutation

In [28], the selection matrix W is randomly generated, which is not the case in the present paper. The selection matrix in this method is modified and given by the expression (4.14)

$$W = \left[W_{ij} \right]_{m \times 1} = W_{(i,j)} = W_{ij}((i,j) \in \{1, 2, \ldots, m, 1, 2, \ldots, m\} \times \{1\}), w_{ij} \in \{0, 1\} \tag{4.14}$$

Modification refers to the matrix W elements which are not of significance for the specified fitness function, that is, which belong to non-priority columns and define the sequence of the gene selection. Namely, such elements equal 0 because they are not the most represented, or taking them into account would lead to a new useless probe configuration.

The current 2-point method was employed for crossover, and current Boolean method was used for mutation. Both methods are described [28].

Modified algorithm was tested for the following parameters:

- $N = m = 16$—population size or the number of features for inspection,
- $G = 5$—genome size,
- $S = 10$—tournament size,
- $G = 100$—number of used genes and
- $p_{per} = 0.02$—value of permutation probability.

4.4.4 Conclusion Remarks

The research carried out in this chapter furnishes the development of a new intelligent approach of probe configuration and setup planning for inspection of prismatic measurement parts on a CMM in order to reduce measurement time and increase the planning process autonomy through minimum human involvement in the process of PMP setups analysis and measuring probes configuration.

Research is directed, in the narrow sense, to obtaining the optimal number of setups and probe configuration for each of the PMP setups on a CMM. In a broader sense, the approach combines complex analyses contained in the developed inspection planning model such as the mathematical model of prismatic parts setup, the measurement sensor accessibility analysis, the probe configuration model, optimal PMP setup and probe configuration. The developed prismatic parts setup model and the probe configuration model, with the application of the GA algorithm at the output, yield optimal number of PMP setups and optimal probe configuration for the cases of using probe star and probe head.

References

1. Stojadinovic S, Majstorovic V, Durakbasa N, Sibalija T (2016) Towards an intelligent approach for CMM inspection planning of prismatic parts. Measurement 92:326–339
2. Stojadinovic S, Majstorovic V, Durakbasa N, Sibalija T (2016) Ants colony optimization of the measuring path of prismatic parts on a CMM. Metrol Measur Syst 23(1):119–132
3. Stojadinovic S, Majstorovic V (2014) Developing engineering ontology for domain coordinate metrology. FME Trans 42(3):249–255
4. Zhao Y, Xu X, Kramer T, Proctor F, Horst J (2011) Dimensional metrology interoperability and standardization in manufacturing systems. Comput Stand Interfaces 33(6):541–555
5. ElMaraghy HA, Gu PH (1987) Expert system for inspection planning. Ann CIRP 36(1):85–89
6. Hussien AH, Youssefy MA, Shoukryz KM (2012) Automated inspection planning system for CMMs. In: Proceedings of the international conference on engineering and technology. IEEE, Cairo, pp 1–6

7. Limaiem A, ElMaraghy AH (1998) Automatic path planning for coordinate measuring machine. In: Proceedings of the 1998 IEEE, international conference on robotics and automation, Leuven, Belgium, pp 887–892
8. Zhao H, Kruth JP, Gestel NV, Boeckmans B, Bleys P (2012) Automated dimensional inspection planning using the combination of laser scanner and tactile probe. Measurement 45:1057–1066
9. Ravishankar S, Dutt HNV, Gurumoorthy B (2010) Automated inspection of aircraft parts using a modified ICP algorithm. Int J Adv Manuf Technol 46:227–236
10. Chang HC, Lin AC (2010) Automatic inspection of turbine blades using 5-axis coordinate measurement machine. Int J Comput Integr Manuf 23(12):1071–1081
11. Chang HC, Lin AC (2011) Five-axis automated measurement by coordinate measuring machine. Int J Adv Manuf Technol 55:657–673
12. Yau HT, Menq CH (2005) Automated CMM path planning for dimensional inspection of dies and molds having complex surface. Int J Mach Tools Manuf 35(6):861–876
13. Zhang SG, Ajmal A, Wootton J, Chisholm A (2000) A feature based inspection process planning system for co-ordinate measuring machine (CMM). J Mater Process Technol 107:111–118
14. Germani M, Mandorli F, Mengoni M, Raffaeli R (2010) CAD-based environment to bridge the gap between product design and tolerance control. Precis Eng 34:7–15
15. Roy U, Xu Y, Wang L (1994) Development of an intelligent inspection planning system in an object oriented programming environment. Comput Integr Manuf Syst 7(4):240–246
16. Stojadinovic S, Majstorović V (2012) Towards the development of feature—based ontology for inspection planning system on CMM. J Mach Eng 12(1):89–98
17. Ziemian CW, Medeiros DJ (1997) Automated feature accessibility for inspection on a coordinate measuring machine. Int J Prod Res 35(10):2839–2856
18. Chiang YM, Chen FL (1999) CMM probing accessibility in a single slot. Int J Adv Manuf Technol 15:261–267
19. Spitz NS, Spyridi JA, Requicha GAA (1999) Accessibility analysis for planning of dimensional inspection with coordinate measuring machines. IEEE Trans Robot Autom 15 (4):714–722
20. Alvarez JB, Fernandez P, Rico CJ, Mateos S, Suarez MC (2008) Accessibility analysis for automatic inspection in CMMs by using bounding volume hierarchies. Int J Prod Res 46 (20):5797–5826
21. Rico JC, Valino G, Mateous S, Cuesta E, Suarez CM (2002) Accessibility analysis for star probes in automatic inspection of rotational parts. Int J Prod Res 40(6):1493–1523
22. Limaiem A, Maraghy HE (1997) A general method for analysing the accessibility of features using concentric spherical shells. Int J Adv Manuf Technol 13:101–108
23. Jackman J, Park K-D (1998) Probe orientation for coordinate measuring machine systems using design models. Robot Comput Integr Manuf 14:229–236
24. Ziemian W, Medeiros JD (1998) Automating probe selection and part setup planning for inspection on a coordinate measuring machine. Int J Comput Integr Manuf 11(5):448–460
25. Lai JY, Chen KJ (2007) Localization of parts with irregular shape for CMM inspection. Int J Adv Manuf Technol 32:1188–1200
26. Lu GC, Morton D, Wu HM, Myler P (1999) Genetic algorithm modelling and solution of inspection path planning on a coordinate measuring machine (CMM). Int J Adv Manuf Technol 15:409–416
27. Liangsheng Q, Guanhua X, Guohua W (1998) Optimization of the measuring path on a coordinate measuring machine using genetic algorithms. Measurement 23:59–170
28. Rice O, Nyman R (2013) Efficiently vectorized code for population based optimization algorithms. UCL Department of Computer Science. Research Note

Chapter 5
Ant Colony Optimisation
of the Measuring Path of PMPs
on a CMM

Abstract This chapter presents optimisation of a measuring probe path in the inspection of prismatic parts on a CMM. The optimisation model is based on the mathematical model that establishes an initial path presented by the set of points, with defined sequence of measuring probe passes without collision, and the solution of the travelling salesman problem (TSP) obtained using ant colony optimisation (ACO). A mathematical model was developed, analysed and presented in Chap. 3, so that a brief reference will be made to it in this chapter in order to relate it to the measuring path optimisation. The problem of finding the shortest measuring probe path in inspection planning of PMP on a CMM is reduced into the TSP solution. TSP is solved by using the techniques of artificial intelligence (AI) such as genetic algorithms (GA), neural networks (NN) and recently the swarm theory (ST). In order to solve TSP, ACO algorithm that aims to find the shortest path of ant colony movement (i.e. the optimised path) is applied. Then, the optimised path is compared with the measuring path obtained by online programming on CMM ZEISS UMM500 and with the measuring path obtained in the CMM module for inspection in software Pro/ENGINEER version Wildfire 4.0 (PTC Creo).

5.1 Introduction

The coordinate measuring machine (CMM) has been recognised as a powerful tool for dimensional and geometric tolerance inspection in the manufacturing industry [1]. Its major characteristic is the versatility of application both in terms of inspection for different types of tolerances and different types of surfaces on machine parts such as gears, turbine blades, rotating parts and prismatic parts.

Demands of a competitive, global market imply reduction of the time needed for inspection on a CMM. The inspection time is directly proportional to the length of a measuring probe path, which is influenced by several factors such as complexity of a measuring part (i.e. workpiece), number of tolerances, and quality and type of tolerances. Reduction of the length of a measuring probe path results in the reduction of the total time needed for inspection, which further reduces inspection

© Springer Nature Switzerland AG 2019
S. M. Stojadinović and V. D. Majstorović, *An Intelligent Inspection Planning System for Prismatic Parts on CMMs*, https://doi.org/10.1007/978-3-030-12807-4_5

costs and total manufacturing time, and thereby more rapid market placement, which is one of the major challenges today.

Ant colony (AC) algorithm, initially proposed by M. Dorigo and explained in detail in [2–5], is one of the swarm intelligence methods and widely accepted metaheuristic optimisation technique.

The travelling salesman problem (TSP) is a basis for ACO application in coordinate metrology. An example of TSP solution is presented in [6]. Besides ACO, other optimisation techniques have been used to obtain the optimal measuring path such as genetic algorithm [7, 8] and neural networks [1], as well as the optimisation algorithm presented in [9]. Data processing for generating the optimal measuring path could be based on an engineering ontology [10–12] and knowledge base [13], where the knowledge base connects geometry and the tolerances of a workpiece. Holistic approach in data definition has been developed [14] to allow the data transfer among different employee groups and departments in an enterprise, with the main goal to improve employees' qualifications and enhance the overall product quality.

Feature-based inspection planning systems that define initial measuring paths are presented in [15–17]. They are used to depict the workpiece geometry and connect it to the tolerances utilising a knowledge base. An optimal path of a measuring sensor is generated for the feature that takes part in tolerance creation. The current status of using geometric tolerances in intelligent manufacturing system is considered in [18].

Automatic inspection planning concepts are presented in [19, 20], and they could serve as a basis for development and implementation of optimisation models. The state of the art in inspection planning is given in [21], and direction of the production metrology development within intelligent manufacturing is shown in [22].

This chapter presents one of the first applications of ACO in coordinate metrology that aims to optimise the measuring path in inspection of prismatic measurement parts (PMPs) on CMM. The main advantage of ACO in solving the observed problem is simple implementation based on the developed mathematical model that reduces the problem to TSP. In particular, inspection of prismatic part with high geometric complexity and metrological diversity, in low volume and individual production, is considered. Besides, the automatic generation of the optimal measuring path based on CAD model also presents an advantage in comparison with the existing optimisation methods from the literature. The developed model could be also applied on hybrid coordinate measuring systems [23], as well as on virtual coordinate measuring machines [24], in order to minimise inspection time and cost.

The rest of the chapter is organised as follows. First, the data model is presented; then the mathematical model and ACO model are depicted, followed by the optimisation results and concluding section. The basis for ACO model is data taken from IGES file of the PMP CAD model and mathematical model that gives a set of points used to find the shortest distance or the optimal path of a measuring probe. Then, as above mentioned, the optimal path is compared with the path obtained by online programming on CMM ZEISS UMM500 and with the path generated in

Pro/ENGINEER, using the same parameters. A prismatic part is produced in order to perform this experiment, i.e. to compare the ACO-based optimal path with the paths obtained by the other two methods.

5.2   Data Model

A model of data needed to apply ACO and to perform comparison with the other two methods is composed of three parts, as presented in Fig. 5.1. The three parts are as follows:

- The first part implies data needed to obtain a measuring path as a result of online CMM programming. This set of data, on one hand, involves the data related to

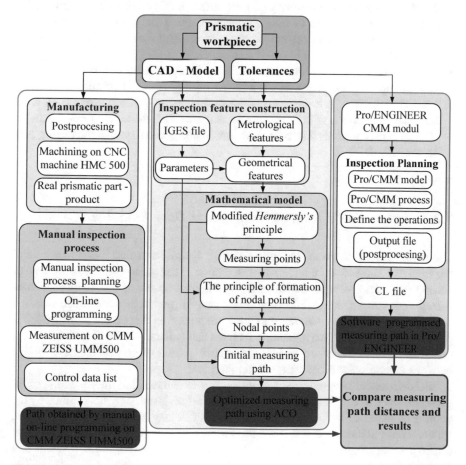

Fig. 5.1 Data model [25]

the design of technological processes for both parts, postprocessing and G-code generation for CNC LOLA HMC500 for previously 3D modelled parts in software Autodesk Inventor Professional 2011, as well as the machining process itself with tools available at the Institute of Machine Tools, Faculty of Mechanical Engineering in Belgrade. On the other hand, it includes data on the manual inspection process planning, manual programming of the machine, measuring process on CMM ZEISS UMM500 and the control data list. Data on the manual inspection process planning are contained in files and are standardised for the laboratory where measurements were performed (High-Precision Measurement Room—Nanometrology Laboratory, TU Wien). The files contain data on the coordinate systems for measurements, configuration of the measuring probes, labels of features involved in creating tolerances, and thereby in the inspection process, etc. Data on the manual programming are those contained inside the control unit and are obtained by memorising the measuring probe path that is initiated by activating the corresponding axis handles. Data on the measuring process refer, in this context, to the measuring path but not to the report on measurements analysed in the previous chapter. The measuring path is represented in the point-to-point form and is found in the control data list.

- The second part refers to the data needed to obtain a set of points or initial path and to apply ACO. These data are based on the model of primitives and represent the result of research carried out in this doctoral dissertation. They serve to make comparison with other data so as to establish the level of advancement.
- The third part presents data needed for automatic generation of a measuring path in Pro/ENGINEER.

5.2.1   Inspection Feature Construction Data Model (IFC)

The notion feature is associated with defining the collision-free measuring path at the level of a single feature and its representation in the point-to-point form. One or several geometric features constitute a metrological primitive that is directly connected with some form of tolerance. For example, parallelism tolerance between two planes contains two features, and these are planes. In that case, the metrological primitive is defined by two geometric features and a datum on the distance between these two planes. In other words, the metrological feature represents a bridge between tolerances and geometry of the features.

IFC data model is based on the basic geometric features and their parameters, as presented in [13] and Chap. 3. Each geometric feature is uniquely determined by the coordinate system O_F, X_F, Y_F, Z_F and the set of belonging parameters. As above stated, these parameters could be of the following types: diameter (D, D_1), height (H, H_1), width (a), length (b), normal vector of a primitive ($\mathbf{n}$), fullness vector of a primitive ($\mathbf{n_p}$). The normal vector $\mathbf{n}$ defines orientation of a feature in a space. The fullness parameter is determined by the unit vector of X-axis of a primitive. The fullness vector $\mathbf{n}_p = \begin{bmatrix} -1 & 0 & 0 \end{bmatrix}$ defines a full primitive, and the vector $\mathbf{n}_p = \begin{bmatrix} 1 & 0 & 0 \end{bmatrix}$ defines an empty primitive.

The fullness parameter and the normal vector define the direction of a measuring probe access in generating the measuring path. Using mentioned parameters, as above stated, each feature is uniquely defined with respect to its coordinate system, O_F, X_F, Y_F, Z_F. With respect to the measurement workpiece coordinate system O_W, X_W, Y_W, Z_W, a feature is determined by the matrix Eq. 4.4 and with respect to the coordinate system of a CMM 4.6 (Chap. 4).

5.2.2   Workpiece Data Model and Processing

In order to generate a workpiece for the measurement on a CMM and for the development of online measuring path, as described in detail in Chap. 3, the processing of a prismatic part was performed on a four-axis ($X'\,Y\,Z\,B'$) CNC machining centre (horizontal milling and drilling machine, type 'LOLA HMC500'), as presented in Fig. 5.2.

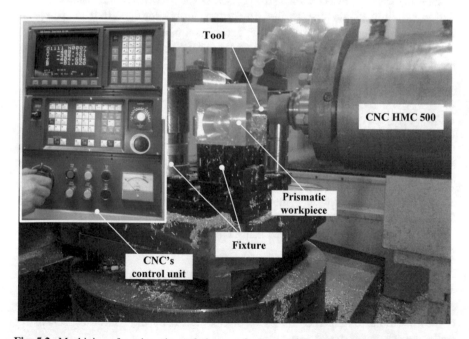

Fig. 5.2 Machining of a prismatic workpiece on four-axis CNC machining centre [25]

Characteristics of a workpiece are as follows:

- Dimensions are 95 × 95 × 95 mm.
- Surface finish quality is N7.
- Material is aluminium.

The following tools were used:

- end mills,
- toolholders for internal cylindrical surfaces and
- spherical milling cutter for hemisphere surfaces.

The generated prismatic workpiece contains all geometric features enclosed in IFC, which are needed to test the ACO-based model.

5.2.3 Online Programmed Measuring Path on a CMM

After referring to the machining process, it is necessary to address in brief the experimental setup of the measuring process on a CMM.

The experimental installation for measurement of the produced PMP, i.e. workpiece, on a CMM is shown in Fig. 5.3. Measurement of the workpiece was performed in a single clamp; the measuring probe configuration is shown in Fig. 5.3; the fixture tool used for PMP fixture is a hand vice.

The experiment was performed on CMM ZEISS UMM500, whose main technical characteristics are:

- Number of axis equals 3 (X, Y, Z).
- Measuring range is 500 × 200 × 300 mm.
- MPE is 0.4 + L/600 µm.
- Assurance is 0.2 µm.

The inspection process is composed of the preparation process and the measuring process. The preparation process involves:

1. setting up the workpiece, with the analysis of fixture tools and accessories,
2. configuration of measuring probes,
3. calibration of measuring probes using calibration sphere and
4. alignment of PMP.

In the measuring process, coordinates of online measuring path are taken from the data list (i.e. STEUERDATENLISTE ZEISS UMESS) for the workpiece. In addition to other data (e.g. measuring regimes, velocity and acceleration), the control list contains the coordinates of points passed by the measuring sensor for online programmed measuring path.

The coordinates of the measuring path points were taken from this data list records, and online, point-to-point measuring path was formed in order to make

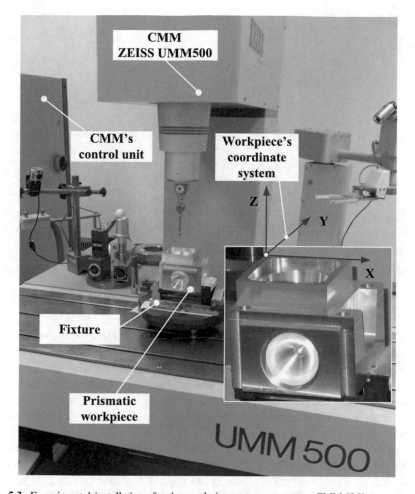

Fig. 5.3 Experimental installations for the workpiece measurement on CMM [25]

comparison with other paths. The deployed online programming method in this subsection is not a novelty, but it is used to obtain one form of a measuring path as a basis for comparison with other measuring paths, as presented by the results for comparison.

5.2.4 Software Programmed Measuring Path in Pro/ENGINEER®

In order to obtain a measuring path, a module 'Manufacturing' and its submodule 'CMM' in Pro/ENGINEER® (version Wildfire 4.0) were used.

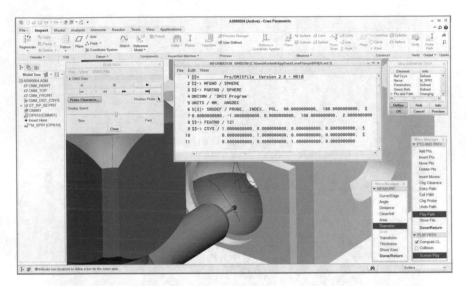

Fig. 5.4 Inspection of a hemisphere in Pro/ENGINEER, with belonging CL file [25]

The coordinate system of a workpiece during the inspection corresponds to the workpiece coordinate system used for the inspection on a CMM. Figure 5.4 shows the measuring path for the inspection of a truncated hemisphere diameter, as well as a part of the generated CL file. The CL file is generated by software as the output report containing the data on the measuring sensor motion. The coordinates of points passed by the measuring sensor are taken from the CL file, and then the software programmed measuring path is formed.

To provide a valid comparison between the optimal measuring path and the software-generated measuring path, it is necessary to appropriately tune several parameters in Pro/ENGINEER, such as APPROACH_DISTANCE, PULLOUT_DIST, MEAS_APPR_DIST and MEAS_PULLOUT_DIST. Values of these parameters must be in accordance with the parameters of mathematical model $(d_1, d_2, \text{etc.})$. These settings are performed in a dialog window 'Measurement step' in the option 'Parameters'.

5.3   Collision Zones

The mathematical model of the measuring points' distribution, as above mentioned, is based on the modified Hemmersly sequences [26] for calculation of coordinates along two axes of a primitive. The model specifies distribution of two sets of points such as:

1. measuring points and
2. node points.

The coordinates of measuring points are presented in the Cartesian coordinate system and named after its author $P_i(x_i, y_i, z_i)$. The set of node points implies two subsets $P_{i1}(x_{i1}, y_{i1}, z_{i1})$ and $P_{i2}(x_{i2}, y_{i2}, z_{i2})$, where $i = 0, 1, 2, \ldots, (N-1)$. The procedure of defining the sets of these points for three types of characteristic surfaces is presented in Chap. 3.

The sets of points P_{i1} and P_{i2} are defined for each of the geometric features depending on their parameters. For the purpose of optimisation, in the inspection of a feature these sets of points demarcate three zones (Fig. 5.5), as follows:

- zone of potential collision between a measuring probe and PMP or collision zone,
- optimal collision zone,
- zone of unprofitable inspection planning or free collision zone.

Connecting the points $P_{i1}(x_{i1}, y_{i1}, z_{i1})$ to $i = 0, 1, 2, \ldots, (N-1)$ generates a broken line, which demarcates the zone between potential collision and optimal zone, whereas the line formed by connecting $P_{i2}(x_{i2}, y_{i2}, z_{i2})$ to $i = 0, 1, 2, \ldots, (N-1)$ represents a boundary between the optimal zone and the zone of unprofitable inspection planning or free collision zone.

The size of each zone depends on the type of a primitive for inspection, its parameters such as R, R_1, H, H_1 and the value of adopted constants d_1, d_2. Also, depending on the type of a feature the zones can be surface and volume. For example, in a circle the zones are surface, whereas in other primitives they are volume.

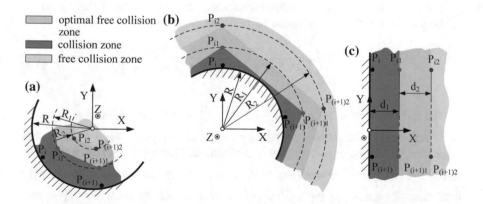

Fig. 5.5 Representation of three collision zones for three characteristic cases of surfaces

5.4 ACO Model

The measuring path could be presented as a set of points (defined by the mathematical model) that a measuring probe passes through during the inspection of a
primitive. Application of ACO in a coordinate metrology is based on the solution of
TSP, where the set of cities that the salesman should pass through with the shortest
possible path corresponds to the set of points of a minimal measuring path length.
Precisely, the set of cities corresponds to the set of points, and the salesman corresponds to the measuring probe. Since it is necessary to avoid collision between
the workpiece and the measuring probe during measurements on a CMM, the
mathematical model must be developed to present distribution of points for basic
geometric features and their unique description, as has been already done in
Chap. 3.

The development of an optimisation model ACO is also preceded by the
development of an algorithm for distribution of the measuring points and an
algorithm for overcoming collision, presented in [27].

In contrast to [28], where A^* algorithm is used for overcoming obstacles, i.e.
collision, the developed two algorithms in this research give collision-free points
which are used as the input for ACO and generation of optimised path. Hence, the
initial path is a collision-free path. Obstacles such as edges, holes and openings are
overcome at two levels. At the first level, the node points for a single feature are
generated, which ensures collision-free inspection of a feature (e.g. during the hole
inspection). The second level implies generation of the collision-free points during
the transition of a measuring sensor from one primitive to another, which means
overcoming of above-mentioned obstacles, i.e. a measurement workpiece volume.
The major novelty is the way in which the initial path is defined and then optimised.

In [28], the path is not collision-free.

The model is based on the following equation for calculation of the measuring
probe path during the measurement on N measuring points:

$$D_{\text{tot}} = \sum_{i=0}^{N-1} \left(\left| \overrightarrow{P_{i2}P_{i1}} \right| + 2 \cdot \left| \overrightarrow{P_{i1}P_i} \right| + \left| \overrightarrow{P_{i1}P_{(i+1)2}} \right| \right) \tag{5.1}$$

The part of a measuring path $2d_1 = 2 \cdot \left| \overrightarrow{P_{i1}P_i} \right|$, which corresponds to the access
(slow movement of a probe) and return movement of a probe for a collision-free
inspection, should be excluded from consideration. This is justified since the value
$2d_1 = 2 \cdot \left| \overrightarrow{P_{i1}P_i} \right|$ is rather small (up to 10 mm), so it does not significantly affect the
total value of a measuring path. On the other hand, measuring points $P_{i1}(x_{i1}, y_{i1}, z_{i1})$
belong to the zone of potential collision, and therefore, it is necessary to exclude
these points from the optimisation process (due to the strictly perpendicular
approach to the surface they belong to). In this way, the generated part of the
measuring path in the potential collision zone between P_i and P_{i1} is added to the

optimised part of the path (optimal zone), and thus, the total optimal collision-free path is obtained for a single primitive. Therefore, the relation (5.1) is transformed to the following relation:

$$D_{\text{tot}} = K + \sum_{i=0}^{N-1} \left(\left| \overrightarrow{P_{i1}P_{(i+1)2}} \right| \right) \tag{5.2}$$

where $K = N \cdot (2 \cdot d_1 + d_2)$ and $d_2 = \left| \overrightarrow{P_{i2}P_{i1}} \right|$.

The relation (5.2) presents equation of the initial path, and it contains constant and variable part.

Since a measuring probe does not have to pass the variable path $\overrightarrow{P_{i1}P_{(i+1)2}}$ every time and other options could also be considered, the optimisation equation, i.e. ant colony path, is obtained based on the initially defined path, as follows:

$$\min\{D_{\text{tot}}\} = K + \left\{ \sum_{i=0}^{N-1} \left(\min\left\{ \left| \overrightarrow{P_{i1}P_{(i+1)2}} \right| \right\} \vee \min\left\{ \left| \overrightarrow{P_{i1}P_{(i+1)1}} \right| \right\} \right. \right.$$
$$\left. \left. \vee \min\left\{ \left| \overrightarrow{P_{i1+1}P_{(i+1)2}} \right| \right\} \vee \min\left\{ P_{i2}P_{(i+1)1} \right\} \right) \right\} \tag{5.3}$$

In defining the initial path, the measuring path is defined by strictly specifying the sequence of points. To provide enough space for optimisation, the sets of points belonging to the optimisation zone P_{i1} and P_{i2} are allowed random visit or pass of the measuring probe to the mentioned points. It is this fact or mathematical model adjustment that enables application of the optimisation principle based on ant colony. Namely, the principle of this technique is a random tour, in this case, of all

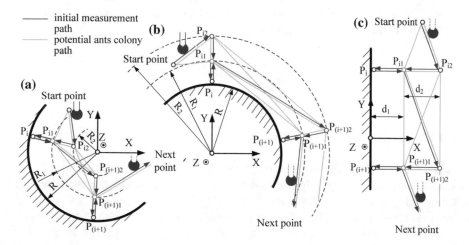

Fig. 5.6 Initial measuring path and ant colony path for concave surface, convex surface and flat surface, respectively [25]

points and pheromone deposition along such path, and then finding the shortest path according to the criterion of the largest amount of deposited pheromones along the travelled path. The path containing the largest amount of pheromones is the shortest path. Number of the random tour cycles and number of the ants in colony may be chosen.

The difference between the initial path and the ant colony path for the observed three cases (concave surface, convex surface and flat surface, respectively) could be noticed in Fig. 5.6.

Red colour marks the initial path with a direction of motion strictly defined by the distribution method, whereas green colour specifies the possible or potential ant colony path. The difference between potential and optimised path is in that the potential path connects each point with each point in the optimal zone, while the optimised path represents only one part of the potential path and it is the part connecting points P_{i1} and P_{i2} by the shortest distance.

5.4.1   Travelling Salesman Problem (TSP)

TSP is the problem of a travelling salesman who starts from the initial city and wants to find the shortest path during his tour of a given set of cities and returns to the initial city, visiting each city only once.

According to [3], TSP can be represented by a complete weighted graph $G = (N, A)$ with N being the set of nodes representing the cities, and A being the set of arcs. Each arc $(i, j) \in A$ is assigned a value (length) d_{ij}, which is the distance between cities i and j, with $i, j \in N$. The goal in TSP is to find a minimum length Hamiltonian circuit of the graph where a Hamiltonian circuit is a closed path visiting each of the $n = |N|$ nodes of G exactly once [3], so that an optimal solution to the TSP is a permutation π of the node indices $\{1, 2, \ldots, n\}$ such that the length $f(\pi)$ is minimal where $f(\pi)$ is given by:

$$f(\pi) = \sum_{i=1}^{n-1} d_{\pi(i)\pi(i+1)} + d_{\pi(n)\pi(1)} \qquad (5.4)$$

5.4.1.1   ACO Algorithm for TSP

ACO algorithm for a TSP solution can be applied by constructing graph $G = (C, L)$ where the set L is completely connected with the set C identically to the graph for representing TSP, where $C = N$ and $L = A$ (Fig. 5.7), and in that case the set of the problem states is matched by the set of all possible partial paths and limitations Ω, which ensures that ants pass and construct only achievable paths that correspond to permutations of cities' indices. Per the length of the constructed paths the ant colony deposit pheromones and create so-called pheromone trail. The pheromone

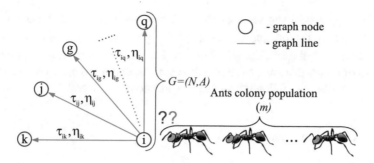

Fig. 5.7 Simplified representation of ant colony motion inside graph G

trails are connected with the graph branches via the function τ_{ij} which represents the tendency to visit city j directly after visiting city i.

A new parameter is heuristic information that is calculated as $\eta_{ij} = 1/d_{ij}$ and represents heuristic tendency to leave the city directly for the city and is inversely proportional to the distance between two cities. The pheromone trail is represented in the form of a pheromone matrix with elements τ_{ij}. Analogously, using heuristic matrix whose elements are η_{ij}, heuristic information is shown.

For simplicity, the ant colony path is constructed in three steps:

1. Applying a specified criterion, choose the initial city where the colony will be positioned.
2. Using τ_{ij} and η_{ij}, construct possible path by iterative addition of cities not already visited by the colony.
3. Return to the initial city.

After all ants in the colony have visited all cities, they can start laying pheromones in the next tour cycle. After each cycle, it is necessary to refresh or re-register the amount of deposited pheromones.

The shortest path contains the largest amount of pheromones.

5.4.1.2 Tour Construction

The ant system (AS) technique is used for the tour construction, where m artificial ants concurrently build a tour of the TSP. Ants randomly visit cities, i.e. in this case these are points $P_{i1}(x_{i1}, y_{i1}, z_{i1})$ and $P_{i2}(x_{i2}, y_{i2}, z_{i2})$ defined by the mathematical model. At each construction step, ant k applies a probabilistic action choice rule, called random proportional rule, to decide which point to visit next. In particular, the probability with which ant k, currently at point i, chooses to go to point j is:

$$p_{ij}^k = \frac{[\tau_{ij}]^\alpha [\eta_{ij}]^\beta}{\sum_{l \in N_i^k} [\tau_{il}]^\alpha [\eta_{il}]^\beta}, \quad if \; j \in N_i^k \tag{5.5}$$

where $\eta_{ij} = \frac{1}{d_{ij}}$ is heuristic value that is available a priori; α and β are parameters which determine the relative influence of the pheromone trail and the heuristic information; N_i^k is the feasible neighbourhood of ants k when being at point i, that is, the set of points that ant k has not visited yet. According to the recommendations for AS technique, the following values are adopted: $\alpha = 1$ and $\beta = 5$.

5.4.1.3 Update of Pheromone Trails

After all the ants have constructed their tours, the pheromone trails are updated. This is done by first lowering the pheromone value on all arcs by a constant factor, and then adding pheromone on the arcs the ants have crossed in their tours. Pheromone evaporation is implemented by

$$\tau_{ij} \leftarrow (1 - \rho)\tau_{ij}, \quad \forall (i,j) \in L \tag{5.6}$$

where $\rho = 0.1$ is pheromone evaporation rate. The parameter ρ is limited to $0 < \rho \leq 1$ and used to avoid unlimited accumulation of the pheromone trails, and it enables the algorithm to 'forget' bad decisions previously taken. After evaporation, all ants deposit pheromone on the arcs they have crossed in their tour:

$$\tau_{ij} \leftarrow \tau_{ij} + \sum_{k=1}^{m} \Delta\tau_{ij}^k, \quad \forall (i,j) \in L \tag{5.7}$$

where $\Delta\tau_{ij}^k$ is the amount of phenomena ant k deposits on the arcs it has visited. It is defined as follows:

$$\Delta\tau_{ij}^k = \left\{ \begin{array}{ll} 1/C^k, & \text{if arc } (i,j) \text{ belongs to } T^k \\ 0, & \text{otherwise} \end{array} \right\} \tag{5.8}$$

where C^k is length of the tour T^k built by the k-ant, computed as the sum of the lengths of the arcs belonging to T^k.

5.5 Optimisation Results

Results of the optimisation could be seen by comparing the length of three observed measuring paths, as depicted in Fig. 5.8, where the online programmed path is shown using blue colour, the software-generated path is presented in green colour, and the optimised path is given using red colour.

The basis for comparison is geometric features that most commonly take part in tolerance creation such as plane, circle, truncated hemisphere, cylinder and truncated cone. Comparison did not include the path generated in transition from one

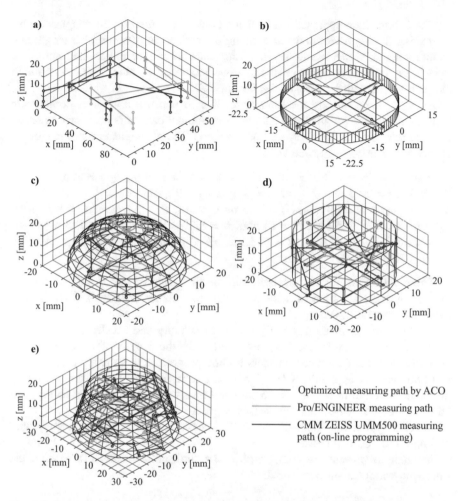

Fig. 5.8 Results and comparison of three obtained measuring paths for geometric primitives: **a** plane, **b** circle, **c** truncated hemisphere, **d** cylinder, **e** truncated cone

feature to another for tolerances whose inspection requires control of two primitives. Namely, these paths are in a function of workpiece dimensions and mutual position and orientation of both features for inspection. Since in all three cases inspection is performed for the same two workpieces, dimensions of the workpieces and mutual position and orientation remain the same, or the paths are approximately equal. So, it is considered that dimensions of a feature have major influence and that the path generated in transition from one feature to another represents the shortest distance between features in the collision-free zone. In this way, comparison is reduced to the comparison between paths generated during inspection for one or several primitives.

The optimisation approach via collision zones, presented in this chapter, is also interesting from the viewpoint of defining dependencies between path length and feature volume, which would be one of the research orientations in the future. It is to be expected that, compared with other two methods, to obtain some reduction in the path length with increase of the feature volume.

As it has been already mentioned, in order to perform a credible comparison between the online programmed path and the optimal path, it is necessary to set up the parameters in the same way for both methods. The parameters of features for the observed prismatic workpiece were:

1. plane: a = 95 mm, b = 59 mm, N = 4, d_1 = 2.6666 mm, d_2 = 8 mm,
2. circle: R = 22.5 mm, N = 4, d_1 = 3 mm, d_2 = 9 mm,
3. hemisphere: R = 20 mm, H_1 = 0.65 mm, N = 8, d_1 = 1.6 mm, d_2 = 8 mm,
4. cylinder: R = 17.5 mm, H = 20 mm, N = 8, d_1 = 1.4 mm, d_2 = 7 mm,
5. truncated   hemisphere: R = 27 mm, H = 76 mm, H_1 = 56 mm, N = 12, d_1 = 2.16 mm, d_2 = 10.8 mm.

These values are selected according to the limitation of the manufacturing and measuring resources used in this experiment (i.e. characteristics of CNC, CMM, tools, fixtures, probes).

Analyses of the measuring regime such as velocity and acceleration are disregarded in this research. Analysis involved only the length of measured paths generated in three mentioned manners (online programmed, software programmed and optimised by applying ACO). The main reason for disregard is the possibility provided to define in advance the mentioned measuring regimes to be identical in all three cases of generating the measuring path.

In ACO application, to obtain the optimal ant colony path was used colony with 500 ants and 100 iterations.

In order to present the comparison and optimisation results (Table 5.1), the following variables are considered:

Table 5.1 Comparison of the measuring path lengths obtained by three methods

Results	Features				
	Plane	Circle	Truncated hemisphere	Cylinder	Truncated cone
d_m (mm)	203.3896	154.3540	244.8584	290.9837	507.0366
d_s (mm)	202.6522	126.4417	183.6755	228.9870	440.1399
d_o (mm)	159.4604	110.2462	159.0962	172.2142	307.6091
$p_s = d_o/d_s$ (%)	78.68	87.19	86.61	75.20	69.88
$p_m = d_o/d_m$ (%)	78.40	71.42	64.97	59.18	60.66
$I_s = 100 - p_s$ (%)	21.32	12.81	13.39	24.80	30.12
$I_m = 100 - p_m$ (%)	21.60	28.58	35.03	40.82	39.34

- d_m is the measuring path length obtained by online programming on CMM ZEISS UMM500.
- d_s is the measuring path length obtained in Pro/ENGINEER.
- d_o is the optimised measuring path length obtained by ACO.
- I_s is the improvement (in percentages) of the measuring path length obtained by ACO in comparison to the path obtained in Pro/ENGINEER.
- I_m is the improvement (in percentages) of the measuring path length obtained by ACO in comparison to the path obtained by online programming on CMM.

Realistic comparison ensures the starting point of identical measuring regimes in all three cases, which is adjustable, and parameters (d_1, d_2, APPROACH_DISTANCE, PULLOUT_DIST, MEAS_APPR_DIST, MEAS_PULLOUT_DIST) adjusted to be identical in all three cases.

The length of the path generated by online programming (d_m) is taken from the control data list on CMM ZEISS UMM500, the software measuring path length (d_s) is taken from the corresponding CL file generated in Pro/ENGINEER, while the optimised path length (d_o) is the shortest path obtained by the ACO model.

The software path is obtained based on the analysis of CL file, which shows points used for the measuring path movement. Table 5.1 shows the values of measuring path lengths obtained by three methods (d_m, d_s and d_o), as well as the result of comparison, i.e. improvements (values I_s and I_m). For the observed features, ratios of the measuring path lengths d_o/d_s are less than 87.19%, which presents reduction of the measuring path obtained by ACO at least by 12.81%. Ratios of the measuring path lengths d_o/d_m are less than 78.40%, presenting reduction of the measuring path at least by 21.60%.

As it has been known in coordinate metrology, the length of a measuring path is directly proportional to the measuring time. Therefore, the obtained lengths of optimised measuring paths directly reduce the total time of inspection for a prismatic workpiece on CMM. The advantage of the proposed model is specially highlighted when there is large number of tolerances in a workpiece and in case of a high geometric complexity of a prismatic workpiece, for medium and high quality of tolerances precision.

5.6   Concluding Remarks

This chapter presents AC-based optimisation of a measuring path in prismatic part inspection on a CMM. The optimisation is based on the previously developed mathematical model for the distribution of measuring and node points of a measuring path, and on the solution of TSP obtained by ACO.

The optimal measuring path is compared with the online programmed measuring path on a CMM and with the automatically generated measuring path in software Pro/ENGINEER (CMM module), for the observed prismatic workpiece. Comparison between the optimal path obtained by ACO and the online programmed path shows improvement, i.e. reduction of a measuring path by more than 20%. The optimal path obtained by ACO is over 10% shorter than the path obtained by Pro/ENGINEER, with the same parameters setting used for both methods.

Besides the mentioned results, the advantage of ACO is a simple implementation based on the developed mathematical model which transforms measuring path to the set of points and converts optimisation problem to TSP. The output of AC-based optimisation is the optimised point-to-point measuring path for measurement of the basic geometric primitives. Similarly to CNC programming (such as G-code), the obtained output could be used for offline programming on a CMM, where the optimised path is given in point-to-point form.

The presented research is conducted within wider research in the development of an intelligent system for inspection planning of PMPs that aims to address main industrial demands such as high geometric variability of modern products and fast delivery, i.e. time to market. The overall goals are:

- to reduce the total manufacturing time, where the proposed research contributes in the reduction of inspection time, and
- to improve the inspection quality through automation of the activities that are usually performed by inspection planner.

Since this model is developed only for the basic geometric primitives, the main limitation of the presented research refers to its usage only for prismatic parts, i.e. the proposed model cannot be used for the inspection of free-form surfaces.

References

1. Hwang CY, Tsai CY, Chang CA (2004) Efficient inspection planning for coordinate measuring machines. Int J Adv Manuf Technol 23:732–742
2. Dorigo M, Gambardella LM (1997) Ant colonies for the travelling salesman problem. BioSystems 43:73–81
3. Dorigo M, Stützle T (2004) Ant colony optimization. The MIT Press Cambridge, Massachusetts London, England
4. Blum C (2005) Ant colony optimization: introduction and recent trends. Phys Life Rev 2:353–373
5. Dorigo M, Blum C (2005) Ant colony optimization theory: a survey. Theoret Comput Sci 344:243–278
6. Cheng TF, Chun TW, Ching TC (2004) A new hybrid heuristic approach for solving large traveling salesman problem. Inf Sci 166:67–81
7. Lu GC, Morton D, Wu HM, Myler P (1999) Genetic algorithm modelling and solution of inspection path planning on a coordinate measuring machine (CMM). Int J Adv Manuf Technol 15:409–416
8. Liangsheng Q, Guanhua X, Guohua W (1998) Optimization of the measuring path on a coordinate measuring machine using genetic algorithms. Measurement 23:159–170
9. Moroni G, Petro S (2013) Inspection strategies and multiple geometric tolerances. Proc. CIRP 10:54–60

10. Pellitero MS, Barreiro J, Cuesta E, Alvarez JB (2011) A new process-based ontology for KBE system implementation: application to inspection process planning. Int J Adv Manuf Technol 57:325–339

11. Stojadinovic S, Majstorovic V (2012) Towards the development of feature–based ontology for inspection planning system on CMM. J Mach Eng 12(1):89–98

12. Stojadinovic S, Majstorovic V (2014) Developing engineering ontology for domain coordinate metrology. FME Trans 42(3):249–255

13. Majstorovic V, Stojadinovic S, Sibalija T (2015) Development of a knowledge base for the planning of prismatic parts inspection on CMM. Acta IMEKO 4(2):10–17

14. Weckenmann A, Werner T (2010) Holistic qualification in manufacturing metrology by enhancing knowledge exchange among different user groups. Metrol Meas Syst 17(1):17–26

15. Myeong CW, Honghee L, Gil YS, Jinhwa C (2005) A feature–based inspection planning system for coordinate measuring machines. Int J Adv Manuf Technol 26:1078–1087

16. Kramer TR, Huang H, Messina E, Proctor FM, Scott H (2001) A feature – based inspection and machining system. Comput Aided Des 33(9):653–669

17. Kamrani A, Nasr EA, Ahmari AA, Abdulhameed O, Mian SH (2014) Feature-based design approach for integrated CAD and computer-aided inspection planning. Int J Adv Manuf Technol 76(9–12):2159–2183

18. Lemu HG (2014) Current status and challenges of using geometric tolerance information in intelligent manufacturing systems. Adv Manuf 2:13–21

19. Yuewei B, Shuangyu W, Kai L, Xiaogang W (2010) A strategy to automatically planning measuring path with CMM offline. In: International Conference on Mechanic Automation and Control Engineering (MACE), Wuhan, China, pp 3064–3067

20. Hussicn HA, Youssefy AM, Shoukryz MK (2012) Automated inspection planning system for CMMs. In: Proceedings of International Conference on Engineering and Technology, Cairo, Egypt, pp 1–6

21. Zhao F, Xu X, Xie SQ (2009) Computer–aided inspection planning—the state of the art. Comput Ind 60(7):453–466

22. Osanna HP (1997) Intelligent production metrology—a powerful tool for intelligent manufacturing. e&iElektrotechnik und Informationstechnik 114:162–168

23. Sladek J, Sitnik R, Kupiec M, Błaszczyk P (2010) The hybrid coordinate measurement system as a response to industrial requirements. Metrol Meas Syst 17(1):109–118

24. Sładek J, Gąska A, Olszewska M, Kupiec R, Krawczyk M (2013) Virtual coordinate measuring machine built using laser tracer system and spherical standard. Metrol Meas Syst 20(1):77–86

25. Stojadinovic S, Majstorovic V, Durakbasa N, Sibalija T (2016) Ants colony optimization of the measuring path of prismatic parts on a CMM. Metrol. Measur Syst 23(1):119–132

26. Lee G, Mou J, Shen Y (1997) Sampling strategy design for dimensional measurement of geometric features using coordinate measuring machine. Int J Mach Tools Manufact 37 (7):917–934

27. Stojadinovic S, Majstorovic V (2015) A feature – based path planning for inspection prismatic parts on CMM, XXI IMEKO World Congress "Measurement in Research and Industry. Prague, Czech Republic, IMEKO, pp 1551–1556

28. Schmitt R, Zheng H, Zhao X, Konig N, Coelho RR (2009) Application of ant colony optimization to Inspection Planning. In: International conference on computational intelligence for measurement systems and applications pp 71–75. IEEE, Hong Kong, China

Chapter 6
Experiment, Results and Concluding Remarks

Abstract For verification of the developed model of inspection planning for prismatic parts on a CMM and measuring path simulation, to visually inspect the collision between the measuring probe and the workpiece, the programme was first written. Apart from verification and simulation, one of the major goals of the written programme is generation of the measuring protocol and a list of output control data, which are then used in the experimental planning process and as an input for experimental measurements. The simulation output is a measuring protocol for CMM ZEISS UMM500. An experiment was performed on two prismatic parts that have been produced for the purpose of this research. The inspection results show that all tolerances for both parts are within the specified limits. The proposed model presents a novel approach of the intelligent inspection planning. The advantages of this approach imply the reduction of preparation time due to an automatic generation of a measuring protocol, a possibility for the optimisation of measuring probe path, i.e. the reduction of a time needed for the actual measurement and increase the planning process autonomy through minimum human involvement in the process of part setups analysis and measuring probes configuration.

6.1 Measuring Probe Path Simulation

As aforementioned, the aim of PMPs inspection simulation is to visually check the measuring probe path from the viewpoint of collision for a given workpiece and its specified tolerances. It is based on a previously developed model of inspection planning and at output gives a measuring protocol and control data list that contains, among other things, data on measurement point coordinates and interposition points. Simulation was developed using three algorithms as follows:

- algorithm for measurement points distribution,
- algorithm for collision avoidance,
- algorithm for probe path planning.

© Springer Nature Switzerland AG 2019

S. M. Stojadinović and V. D. Majstorović, *An Intelligent Inspection Planning System for Prismatic Parts on CMMs*, https://doi.org/10.1007/978-3-030-12807-4_6

6.1.1   Algorithm for Measurement Point Distribution (AMPD)

As presented in Fig. 6.1, AMPD consists of five steps: S1, S2, S3, S4 and S5.

The step S1 implies loading of the algorithm input parameters. As it has been previously mentioned, one group of these parameters such as a, b, R, R_1, H, H_1, $\mathbf{n_p}$, $\mathbf{n_{pi}}$ and T is taken over from IGES file of PMP CAD model, whereas N—number of measurement points is specified depending on the form and quality of specified tolerances.

The constant d_2 is chosen according to the primitive volume, and then, the constant d_1 is calculated over it. For example, for the cylinder of radius $R = 17.5$ mm and height $H = 17.5$ mm, the formulas for calculating constants are $d_2 = 0.4R$ and $d_1 = 0.2d_2$.

In addition, keywords—plane, circle, cylinder, cone, cone_zar, sphere_zar— are needed for the subroutine call in the step S3 of the algorithm and procedure **KEYWORD**, respectively. Keywords include all geometric features represented by the developed model of inspection planning, and their names correspond to the names of primitives: plane, circle, cylinder, cone, truncated cone, sphere and truncated sphere, respectively.

The constant k used for the next step of the algorithm is calculated in the step S2. This constant is taken over from Hammersley's formula so that in a modified formula it retains the same meaning as in the original formula, depending on the function log base 2 and desired number of measurement points (N).

Fig. 6.1 Procedure for measurement points distribution

Procedure *ADMP*
S1: Read parametars: $N; a; b; R; R_1; H; H_1; \mathbf{n}; \mathbf{n_p}; T; d_2;$ **keyword=input(plane; circle; cylinder, cone; cone_zar; sphere; sphere_zar).**
S2: Calculation: $k = \log_2 N$
S3: Call procedure *KEYWORD*
S4: Homeogenic coordinates: $\left\{ P = \begin{bmatrix} s & t & w & h \end{bmatrix}^{T} \right\}$ $\left\{ P_1 = \begin{bmatrix} s_1 & t_1 & w_1 & h \end{bmatrix}^{T} \right\}$ $\left\{ P_2 = \begin{bmatrix} s_2 & t_2 & w_2 & h \end{bmatrix}^{T} \right\}$
S5: Transformed coordinates: $\left\{ P^{tran} = T \cdot P \right\}$ $\left\{ P_1^{tran} = T \cdot P_1 \right\}$ $\left\{ P_2^{tran} = T \cdot P_2 \right\}$

In the next step S3, the calculation of the point coordinates $P_i(x_i, y_i, z_i)$, $P_{i1}(x_{i1}, y_{i1}, z_{i1})$ and $P_{i2}(x_{i2}, y_{i2}, z_{i2})$ is performed for different features for the following four cases:

- case plane,
- case circle,
- case cylinder, case cone, case cone_zar,
- case sphere, case sphere_zar.

Each of these cases is called over previously mentioned keywords by applying the procedure KEYWORD given in Fig. 6.2.

For the case of the keyword call *case plane,* the vectors $\overrightarrow{PP_{i1}}$ and $\overrightarrow{PP_{i1}}$ (step P1) are formed first, and then, the coordinates of measurement points $P_i(x_i, y_i, z_i)$ are calculated using modified Hammersley's formula and points $P_{i1}(x_{i1}, y_{i1}, z_{i1})$ and $P_{i2}(x_{i2}, y_{i2}, z_{i2})$ (step P2).

For the case of the keywords call *case circle,* the coordinates of measurement points are calculated first, and then, based on a specified vector $\mathbf{n}_p$, it is checked whether the inspection of the circle is internal or external and, based on this, the vector $\mathbf{n}_{pi}$ is formed. The formed vector $\mathbf{n}_{pi}$ is used to form vectors $\overrightarrow{PP_{i1}}$ and $\overrightarrow{PP_{i2}}$ (step P1).

For the case of the keywords call *case cylinder, case cone* and *case cone_zar,* the same case call is performed with different key words. This is the result of identical distribution of the coordinates of point sets $P_{i1}(x_{i1}, y_{i1}, z_{i1})$ and $P_{i2}(x_{i2}, y_{i2}, z_{i2})$, i.e. the usage of the same formulations for the coordinates' calculation. Similar to the previous key word, in these key words, the coordinates of measurement points are calculated first, and afterwards, it is checked whether it is 'external' or 'internal' inspection of given primitives and, based on inspection, the vectors $\overrightarrow{PP_{i1}}$ and $\overrightarrow{PP_{i2}}$ (step P1) are formed.

For keywords *case sphere* and *case sphere_zar,* the procedure is identical to that for the previously mentioned two keywords, the difference being in w_i-coordinate of the point O_i. Namely, due to collision avoidance at the feature level, represented by these key words, the z-coordinate has the value zero and not the value w_i as in the previous case. This alteration allows for the measuring probe to always approach perpendicularly to the measuring surface in the inspection of a feature, such as sphere and truncated sphere, which eliminates the possibility of collision at the level of one of these features.

The step S4 involves the formation of homogeneous coordinates. All three point sets are recorded in a matrix form represented with the help of the system of matrix equations 6.1. This is necessary because the developed method of inspection planning considers primitives from the aspect of position and orientation as well. Transition to homogeneous coordinates allows for transformations between the coordinate system (CS) such as the piecework CS, the measuring sensor CS, and the measuring machine CS, and determination of a primitive's position and orientation relative to them.

Procedure KEYWORD

Switch keyword
  case plane
  P1: Calculation: $\overrightarrow{PP_1} = \mathbf{n} \cdot d_1 \,; \overrightarrow{PP_2} = \mathbf{n} \cdot d_2$

  P2: For $\left(i = 0 : (N-1)\right)\ \{h_i = i + (1-i)\}$

    for $\left(j = 0 : (k-1)\right) \rightarrow$

$$\left\{ \sum_{j=0}^{k-1} \left(\left[\frac{i}{2^j}\right] \text{Mod} 2\right) \cdot 2^{-(j+1)} \right\}$$

    end
    Calculation: s_i, t_i, w_i
    Set $(q = s, t, w)$ and $(\lambda = 1, 2)$
$$\rightarrow \left\{ q_{\lambda i} = \overrightarrow{PP_\lambda} \cdot \vec{i} + s_i \right\}$$
    end
  end
  case circle
  P1: For $\left(i = 0 : (N-1)\right)\ \{h_i = i + (1-i)\}$

    Calculation: s_i, t_i, w_i

$$\text{If } \left(\mathbf{n}_p \cdot \vec{i} = -1\right) \left\{ s_{ni} = \frac{s_i}{R} \right\}$$

$$\left\{ t_{ni} = \frac{t_i}{R} \right\}\left\{ w_{ni} = \frac{w_i}{R} \right\}$$

$$\text{else } \left\{ s_{ni} = -\frac{s_i}{R} \right\}$$

$$\left\{ t_{ni} = -\frac{t_i}{R} \right\}\left\{ w_{ni} = -\frac{w_i}{R} \right\}$$

    end
    Calculation: $\left\{ n_{pi} = [s_{ni} \quad t_{ni} \quad w_{ni}]^T \right\}$
$$\left\{ \overrightarrow{PP_1} = \mathbf{n}_{pi} \cdot d_1 \right\}\left\{ \overrightarrow{PP_2} = \mathbf{n}_{pi} \cdot d_2 \right\}$$
    Set $(q = s, t, w)$ and $(\lambda = 1, 2)$
$$\rightarrow \left\{ q_{\lambda i} = \overrightarrow{PP_\lambda} \cdot \vec{i} + s_i \right\}$$
    end
  end
  case cylinder
  case cone
  case cone_zar
  P1: For $\left(i = 0 : (N-1)\right)\ \{h_i = i + (1-i)\}$

    for $\left(j = 0 : (k-1)\right)$

$$\left\{ \sum_{j=0}^{k-1} \left(\left[\frac{i}{2^j}\right] \text{Mod} 2\right) \cdot 2^{-(j+1)} \right\}$$

    end

Calculation: s_i, t_i, w_i,

$$\left\{ \overrightarrow{O_i} = [0 \quad 0 \quad w_i] \right\}$$

$$\text{If } \left(\mathbf{n}_p \cdot \vec{i} = -1\right)$$

$$\left\{ s_{ni} = \frac{s_i}{R} \right\}\left\{ t_{ni} = \frac{t_i}{R} \right\}\left\{ w_{ni} = 0 \right\}$$

  else

$$\left\{ s_{ni} = -\frac{s_i}{R} \right\}\left\{ t_{ni} = -\frac{t_i}{R} \right\}\left\{ w_{ni} = 0 \right\}$$

  end

$$\left\{ n_{pi} = [s_{ni} \quad t_{ni} \quad w_{ni}]^T \right\}\left\{ \overrightarrow{PP_1} = \mathbf{n}_{pi} \cdot d_1 \right\}$$

$$\left\{ \overrightarrow{PP_2} = \mathbf{n}_{pi} \cdot d_2 \right\}$$

Set $(q = s, t, w)$ and $(\lambda = 1, 2)$

$$\rightarrow \left\{ q_{\lambda i} = \overrightarrow{PP_\lambda} \cdot \vec{i} + s_i \right\}$$

  end
end
case sphere
case sphere_zar
P1: For $\left(i = 0 : (N-1)\right)\ \{h_i = i + (1-i)\}$

  for $\left(j = 0 : (k-1)\right)$

$$\left\{ \sum_{j=0}^{k-1} \left(\left[\frac{i}{2^j}\right] \text{Mod} 2\right) \cdot 2^{-(j+1)} \right\}$$

  end

Calculation: s_i, t_i, w_i,

$$\left\{ \overrightarrow{O_i} = [0 \quad 0 \quad 0] \right\}$$

$$\text{If } \left(\mathbf{n}_p \cdot \vec{i} = -1\right)$$

$$\left\{ s_{ni} = \frac{s_i}{R} \right\}\left\{ t_{ni} = \frac{t_i}{R} \right\}\left\{ w_{ni} = 0 \right\}$$

  else

$$\left\{ s_{ni} = -\frac{s_i}{R} \right\}\left\{ t_{ni} = -\frac{t_i}{R} \right\}\left\{ w_{ni} = 0 \right\}$$

  end

$$\left\{ n_{pi} = [s_{ni} \quad t_{ni} \quad w_{ni}]^T \right\}\left\{ \overrightarrow{PP_1} = \mathbf{n}_{pi} \cdot d_1 \right\}$$

$$\left\{ \overrightarrow{PP_2} = \mathbf{n}_{pi} \cdot d_2 \right\}$$

Set $(q = s, t, w)$ and $(\lambda = 1, 2)$

$$\rightarrow \left\{ q_{\lambda i} = \overrightarrow{PP_\lambda} \cdot \vec{i} + s_i \right\}$$

  end
end

Fig. 6.2 Procedure KEYWORD for measurement point distribution [1]

$$P = \begin{bmatrix} s & t & w & h \end{bmatrix}^T$$
$$P_1 = \begin{bmatrix} s_1 & t_1 & w_1 & h \end{bmatrix}^T \qquad (6.1)$$
$$P_2 = \begin{bmatrix} s_2 & t_2 & w_2 & h \end{bmatrix}^T$$

The homogeneous coordinate record enables to perform multiplication with the matrix T (from the left side), in order to perform their transformations from the features of coordinate system $(O_F X_F Z_F Y_F)$ to the coordinate system of the measuring part $(O_W X_W Z_W Y_W)$, which is performed in the step S5.

6.1.2 Algorithm for Collision Avoidance (ACA)

Algorithm for collision avoidance is based on the previously described principle for collision avoidance. As stated, the principle, and ACA itself as well, is founded on a STL model of representing geometry of PMP, its tolerances, coordinates of the end point $P_{(N_{F1})1}$ for inspection of the preceding feature and coordinates of the start point $P_{(N_{F2})1}$ for inspection of the next feature.

The algorithm is iterative and consists of permanent intersection between the workpiece volume, represented in STL format, and the corresponding line segment until the moment when there is no intersection remaining between the volume and the line segment. The absence of intersection between the line segment and the volume is an indicator of collision absence between the measuring probe and the workpiece. The complete algorithm ACA is shown in Fig. 6.3.

The algorithm input parameters are the value of correction parameter along the z-axis labelled as δ (mm) (step A1), STL file exported based on CAD model of PMP (step A2) and coordinates of the points $P_{(N_{F1})1}$ and $P_{(N_{F2})1}$ (step A3).

By the loading of vertex coordinates of first triangle $\Delta T_1 T_2 T_3$ (step A4), the equation of a plane through three points is formulated (step A5). At the same time, the equation of a line p through the loaded points $P_{(N_{F1})1}$ and $P_{(N_{F2})1}$ is generated (step A6). Now, it is necessary to check whether the intersection point P_j, $j \in \{0, 1, 2, 3, \ldots, q\}$ between the formed plane and line exists (step A7), and whether it belongs to the line segment $\overline{P_{(N_{F1})1} P_{(N_{F2})1}}$ (step A8). If the intersection point P_j does not exist or it does not belong to the line segment $\overline{P_{(N_{F1})1} P_{(N_{F2})1}}$, then the vertex coordinates of the next triangle are loaded (step A9). If the intersection point P_j exists and belongs to the line segment, then it is necessary to check whether the intersection point is located in the surface part that is limited by the triangle (step A10). Further, check is made on conditions of the angles: $\alpha \leq \alpha_1$ and $\beta \leq \beta_1$ and $\gamma \leq \gamma_1$. If one of the mentioned three conditions is not met, then the intersection point does not belong to the surface part limited by the triangle. In this case, a new triangle is loaded (step A9). If all three conditions of the angles are met, that means the point is located in the surface part limited by the triangle $(P_j \in \Delta T_1 T_2 T_3)$, and it is needed to perform correction of z-coordinates of the loaded points using the

Fig. 6.3 Algorithm for collision avoidance in 15 steps [1]

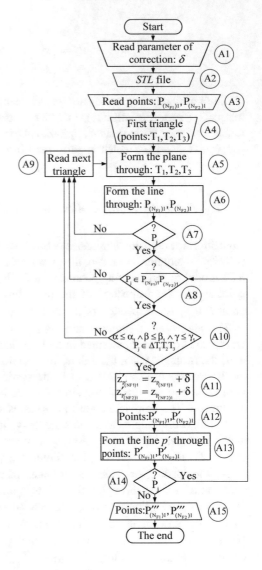

formulas (6.2) in order to avoid the collision between PMP and a measuring probe (step A11).

$$z'_{P_{(NF1)1}} = z_{P_{(NF1)1}} + \delta$$
$$z'_{P_{(NF2)1}} = z_{P_{(NF2)1}} + \delta \tag{6.2}$$

After the mentioned correction, the points $P'_{(N_{F1})1}\left(x_{P_{(NF1)1}} y_{P_{(NF1)1}} z'_{P_{(NF1)1}}\right)$ and $P'_{(N_{F2})1}\left(x_{P_{(NF2)1}} y_{P_{(NF2)1}} z'_{P_{(NF2)1}}\right)$ (step A12) as well as the line p' that contains these points must be formulated (step A13). Then, it is necessary to check whether a new intersection point P_j between p' and the loaded plane exists (step A14). If it exists, then the previously described checking procedure is repeated. If it does not exist, then the corrected points $P'_{(N_{F1})1}$ and $P'_{(N_{F2})1}$ are adopted as collision-free points (step A15). In the example given as an illustration of the collision avoidance principle (Chap. 3), ACA performed three corrections of z-coordinates and the points $P'''_{(N_{F1})1}$ and $P'''_{(N_{F2})1}$ are adopted.

6.1.3 Algorithm for Probe Path Planning (APPP)

APPP is based on AMPD and ACA (Fig. 6.4). It consists of six steps. In the first step, AMPD for feature F1 is loaded and gives the output: $\left(P^{tran}\right)_{F1}$, $\left(P_1^{tran}\right)_{F1}$ and $\left(P_2^{tran}\right)_{F1}$. In the second step, three matrixes are formed: S_{mo}^{F1}, T_{mo}^{F1} and W_{mo}^{F1}, and the arrangements of their elements correspond to the measuring probe movement. The S3 and S4 steps are similar to the first and second step, but they refer to the feature F2.

The outputs from the first four steps are matrixes with precisely defined movement of a measuring probe for F1 and F2, separately. In the fifth step, it is necessary to call ACA and obtain $P'''_{(N_{F1})1}$ and $P'''_{(N_{F2})1}$ to complete the collision-free path for the inspection of features F1 and F2. In the sixth step, the matrix M gives all coordinates of the points, including points from the previous five steps, in the regularly established arrangement.

6.1.4 Generation of the Measuring Protocol and Control Data List

Developed model of the measuring probe path simulation gives at output a measuring protocol and control data list for both test PMPs. An example of a part of the control data list is shown in Fig. 6.5. The measuring protocol is further used for online programming of measuring machine ZEISS UMM500, generating programmes for this machine, measurement process and verification of the developed model for inspection planning.

6.1.5 Experimental Setup

Experimental setup involves all activities necessary for preparation and measuring process performance in the test parts. The activities include the design and

---

Procedure APP

S1: Read **ADMP** for **F1**:

$$\rightarrow \left(P^{tran}\right)_{F1}, \left(P_1^{tran}\right)_{F1}, \left(P_2^{tran}\right)_{F1}$$

S2: Form:

$$S_{mo}^{F1} = \left[\left(s_2\right)_{1j}^{F1} \quad \left(s_1\right)_{1j}^{F1} \quad \left(s\right)_{1j}^{F1} \quad \left(s_1\right)_{1j}^{F1}\right];$$

$$T_{mo}^{F1} = \left[\left(t_2\right)_{2j}^{F1} \quad \left(t_1\right)_{2j}^{F1} \quad \left(t\right)_{2j}^{F1} \quad \left(s_1\right)_{2j}^{F1}\right];$$

$$W_{mo}^{F1} = \left[\left(w_2\right)_{3j}^{F1} \quad \left(w_1\right)_{3j}^{F1} \quad \left(w\right)_{3j}^{F1} \quad \left(w_1\right)_{3j}^{F1}\right], j=1,2,3,...,N$$

S3: Read **ADMP** for **F2**:

$$\rightarrow \left(P^{tran}\right)_{F2}, \left(P_1^{tran}\right)_{F2}, \left(P_2^{tran}\right)_{F2}$$

S4: Form:

$$S_{mo}^{F2} = \left[\left(s_2\right)_{1j}^{F2} \quad \left(s_1\right)_{1j}^{F2} \quad \left(s\right)_{1j}^{F2} \quad \left(s_1\right)_{1j}^{F2}\right];$$

$$T_{mo}^{F2} = \left[\left(t_2\right)_{2j}^{F2} \quad \left(t_1\right)_{2j}^{F2} \quad \left(t\right)_{2j}^{F2} \quad \left(s_1\right)_{2j}^{F2}\right];$$

$$W_{mo}^{F2} = \left[\left(w_2\right)_{3j}^{F2} \quad \left(w_1\right)_{3j}^{F2} \quad \left(w\right)_{3j}^{F2} \quad \left(w_1\right)_{3j}^{F2}\right], j=1,2,3,...,N$$

S5: Read **ACA** for **F1-F2**:

$$\rightarrow P_{(N_{F1})1}''' = \left[s_{F1}''' \quad t_{F1}''' \quad w_{F1}'''\right]; P_{(N_{F2})1}''' = \left[s_{F2}''' \quad t_{F2}''' \quad w_{F2}'''\right]$$

S6: Form the matrix of movement:

$$M = \begin{bmatrix} s_{11} & s_{12} & ... & s_{1\mu} \\ t_{21} & t_{22} & ... & t_{2\mu} \\ w_{31} & w_{32} & ... & w_{2\mu} \end{bmatrix}, \mu = 1,2,3,...,2(j+1),$$

за $\mu = 1,2,3,...,j \Rightarrow s_{1\mu} = s_{1j}^{F1}; t_{2\mu} = t_{2j}^{F1}; w_{3\mu} = w_{3j}^{F1}$,

за $\mu = j+1 \Rightarrow s_{1\mu} = s_{F1}'''; t_{2\mu} = t_{F1}'''; w_{3\mu} = w_{F1}'''; \mu = j+2 \Rightarrow s_{1\mu} = s_{F2}'''; t_{2\mu} = t_{F2}'''; w_{3\mu} = w_{F2}'''$,

за $\mu = (j+2),...,2(j+1) \Rightarrow s_{1\mu} = s_{1j}^{F2}; t_{2\mu} = t_{2j}^{F2}; w_{3j} = w_{3j}^{F2}$

Fig. 6.4 Measuring path planning procedure in six steps [1]

machining of workpieces, calibration of CMM 'ZEISS UMM500' based on arte-facts according to ISO 10360 standard and measurement plan that contains all data for writing a measurement programme.

6.1.6 Design of Prismatic Parts

For verification of the developed inspection planning model, two test PMPs were designed. The drawing of the first simpler test workpiece is shown in Fig. 6.6. The part was designed to contain all geometric primitives embraced by the developed model and a considerable number of standard tolerance types (Table 6.1).

```
================================================================================

                        STEUERDATENLISTE    ZEISS  UMESS

WERKSTUECKNAME:    WP-1par
DATEINAME:  CNC_____53B
STEUERDATENZEILEN:    297              SOLLWERTZEILEN:      0
================================================================================
      |      X      |     Y     |     Z    |         |     |     |     |      |
 NR   |-------------------------------------| Funktion | SKZ | AKZ | PKZ |StKZ | ADR
      |          Dialog           |         |         |     |     |     |      |
 NR   |   Nennmass  | o.Tol | u.Tol |       | Funktion | SKZ | AKZ | PKZ |StKZ |
 NR   |        Bezeichnung         |        | Funktion | SKZ | AKZ | PKZ |StKZ | ADR
================================================================================
   1        -39.3000   0.5000  -0.5000 LFZ SOLLW S2       1     0  9919      0
   2                0   0    0     0 SOLLWERTE SN        0     2  1459      0
   3                 0.0000   0.0000 LFZ SOLLW SN        0     0  9919      0
   4       31491                                     14528     0  8224      2 8224
   5 wp1                          PROTOKOLLKOP          0     8  1610   1650
   6 ZEICHNUNGS NR      |          FZ P-KOPF            0     0  9911      0
   7 AUFTRAGS NR        |          FZ P-KOPF            0     0  9911      0
   8 LIEFERANT/KUNDE    |          FZ P-KOPF            0     0  9911      0
   9 ARBEITSGANG                   FZ P-KOPF            0     0  9911      0
  10 |                             FZ P-KOPF            0     0  9911      0
  11                               LFZ P-KOPF           0     0  9919      0
  12    34.4735   62.7517   21.1390 ZW-POS              0 11110      0   1101

  13                               FLAECHE              0     0  1103   1410
  14    34.9983    5.3066   21.4279 ZW-POS              0 11110      0   1101
  15    13.4146    5.1101   21.4796 ZW-POS              0 11110      0   1101
  16    13.3693    5.0201    3.9978 ANTASTUNG -Z        0 11107      0   1103
  17    13.3897    5.0573   11.0016 ZW-POS              0 11110      0   1101
  18    88.2909    5.7406   10.8213 ZW-POS              0 11110      0   1101
```

Fig. 6.5 Part of the control data list for PMP 1

The drawing of the second more complex test prismatic part is shown in Fig. 6.7. Compared to the test PMP1, the essential difference is in the number of tolerances, elevated tolerance quality and geometric complexity that is more remarkable in the test PMP2. Also, the second part has new types of tolerances to be tested.

Machining of the designed parts is shown in Chap. 5.

6.1.7   Calibration of CMM According to Standard ISO 10360-2

Prior to the inspection process of designed and machined test parts, the accuracy of the machine was checked based on the artefact of domestic production (Fig. 6.8) according to standards ISO 10360-2 [2] and ISO 15530-3 [3].

The artefact was designed and developed at the Vienna University of Technology (TUV). Setting up and clamping the artefact on the machine workbench are performed over three points designated with positions 1, 2 and 3. Calibration conditions include temperature $T = 20\ ^\circ C$, the use of hygienic

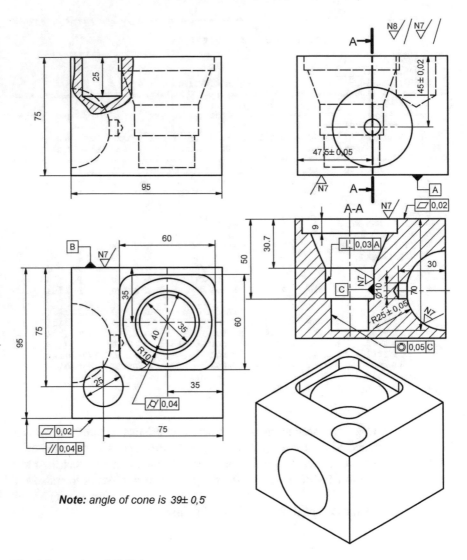

Fig. 6.6 Drawing of PMP 1

equipment such as slippers and gloves, as well as the prescribed distance from the artefact during measurements because of violating the prescribed intensity of the temperature field. On the artefact, it is possible to test the accuracy of the machine for length measurements, diameter, perpendicularity and parallelism. It consists of five calibration spheres internally labelled A, B, C, H and M, diameter $D = 38.1$ mm. Measurements deployed the existing measuring protocol and a programme for the measuring machine. The part of the measurement results repeated 5 times is shown in Fig. 6.9.

Table 6.1 Tested standard type of tolerance

Type of tolerance	Geometric characteristics	Tolerance zone		Tested
		Circle/cylinder	Between two planes	
Form	Straightness	Yes	Yes	No
	Flatness	No	Yes	Yes
	Circularity	Yes	No	Yes
	Cylindricity	Yes	No	Yes
Profile	Profile of a line	–	–	No
	Profile of a surface	–	–	No
Orientation	Parallelism	No	Yes	Yes
	Perpendicularity	No	Yes	Yes
	Angularity	No	Yes	Yes
Location	Position	Yes	Yes	Yes
	Concentricity	Yes	No	Yes
	Symmetry	Yes	Yes	No
Run-out	Circular run-out	–	–	No
	Total run-out	–	–	No

Observing data for distance between spheres B–H (3,551,056 mm), B–C (3,769,612 mm) and B–M (18,848 mm), given by the formula in Fig. 6.9, it can be concluded that standard deviation of measurement results is not higher than 0.2 μm.

Identical conclusion can be drawn about the analysis of results for the diameters of spheres, where the value of standard deviation does not exceed 0.2 μm in any of the cases. Taking into account the dimensions of the test parts, this furthermore implies completely satisfactory dimensional accuracy and the diameter inspection accuracy.

The analysis of the rest of results leads to the same conclusion for perpendicularity and parallelism tolerances.

6.1.8 Inspection Plan for Prismatic Parts

Inspection plan contains all auxiliary data necessary for writing a programme for CMM. For the laboratory where measurements were conducted, the data are kept on a separate internal form. The example of both test parts provides a form with necessary data, Figs. 6.10 and 6.11.

For test PMPs, the CSs are shown, as well as geometric features, whose inspection is planned, and some of the measures like coordinates of the sphere centre, axes position of the cylinders and cones, data and commands for CMM for

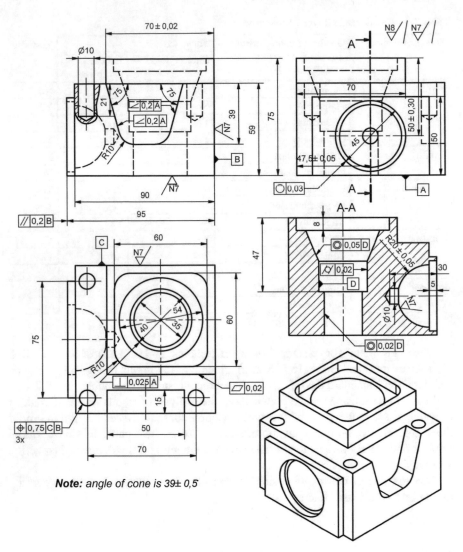

Fig. 6.7 Drawing of PMP 2 [1]

the measuring part alignment are all denoted. Alignment is composed of fixing *z*-axis over plane *A* and then fixing of *x*-axis over plane *D*. The axis *y* and the coordinate origin are formed by the contact at a single point in plane *A*.

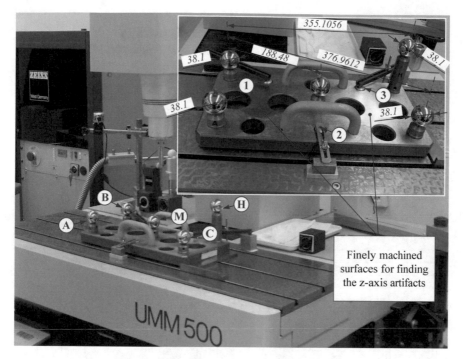

Fig. 6.8 Artefact for calibration of CMM

6.2 Inspection of Prismatic Parts

Inspection of PMPs was performed at the Vienna University of Technology in High Precision Measurement Room—Nanometrology Laboratory. Laboratory is separated mechanically and structurally (the room-in-room concept) by the wall of 30 cm depth and has vibration isolation from the environment. It consists of the laboratory space, where measuring instruments are stored, the control room and anteroom. The space with measuring instruments is kept under controlled influence of the environment on measurements performed with respect to the temperature (temperature conditions, convection and radiation), vibrations, air humidity and pollution. Reliable operation of air conditioners provides continuous supervision of the parameters such as temperature, air humidity, air flow and particles' presence in the laboratory.

Inspection process in our experiment is composed of the preparation process and the measuring process. The preparation process involves:

- setting up the workpiece, with the analysis of fixture tools and accessories,
- configuration of measuring probes,
- calibration of measuring probes using calibration sphere,
- alignment of PMP.

TECHNISCHE UNIVERSITÄT WIEN	**TU** VIENNA		*Prüfprotokoll* *Ermittlung der Meßunsicherheit*					
Prüfmittel: Prüfkörper d. TU			*KMG:*			*Bearbeiter*		*Blatt von*
Adr.-Nr.	*Pos. Bez.*	*Nennmaß*	*1. Messung*	*2.*	*3.*	*4.*	*5. Messung*	*Streuung in μm*
4	KUGEL A	38,1	38,0984	38,0984	38,0983	38,0983	38,0984	0,1
5	KUGEL B	38,1	38,0996	38,0995	38,0994	38,0992	38,0993	0,2
10	KUGEL M	38,1	38,0983	38,0983	38,0983	38,0983	38,0983	0
11	KUGEL C	38,1	38,0988	38,0986	38,0987	38,0987	38,0986	0,1
12	KUGEL H	38,1	38,0988	38,0987	38,0988	38,0987	38,0987	0,1
15	BH RD	355,1056	355,1024	355,1028	355,1027	355,1027	355,1027	0,2
	BH WX		9,7262	9,7262	9,7262	9,7262	9,7262	0
	BH WY		89,9792	89,9793	89,9793	89,9793	89,9793	0
	BH WZ		80,2738	80,2738	80,2738	80,2738	80,2738	0
16	BH X		349,9982	349,9986	349,9985	349,9986	349,9985	0,2
	BH Y		0,1291	0,1282	0,1283	0,1281	0,1282	0,4
	BH Z		59,9912	59,9912	59,9911	59,9909	59,9911	0,1
19	BC RD	376,9612	376,8863	376,8868	376,8867	376,8868	376,8867	0,2
	BC WX		21,7813	21,7814	21,7814	21,7814	21,7814	0
	BC WY		111,7813	111,7814	111,7814	111,7814	111,7814	0
	BC WZ		90,0008	90,0008	90,0008	90,0008	90,0008	0
20	BC X		349,9794	349,9795	349,9795	349,9795	349,9794	0,1
	BC Y		139,849	139,85	139,8498	139,85	139,85	0,4
	BC Z		0,0053	0,0052	0,0053	0,0055	0,0052	0,1
23	BM RD	188,48	188,475	188,4754	188,4751	188,4753	188,4753	0,2
	BM WX		21,7749	21,775	21,775	21,775	21,775	0
	BM WY		111,7749	111,775	111,775	111,775	111,775	0

Fig. 6.9 Part of measurement results for the artefact

The measuring process is based on a measuring protocol obtained as the output from the simulation process and control data list (STEUERDATENLISTE ZEISS UMESS), which is also obtained as the output. The measurement of both parts was done in a single clamp and the measuring probe configuration, as shown in Figs. 6.12 and 6.13. The generated measuring protocol was used for CMM programming. The developed model employs the base of available software UMESS as an evaluation criterion for the measuring process. Experimental installation for the measurement of PMP1 is displayed in Fig. 6.12 and of PMP2 in Fig. 6.13.

The fixture tool used to fix the PMP is the clamp. Another fixture tool used to bring the workpieces to reachable workspace for measuring probes, without their collision with the workbench of CMM, is a flat measuring stand supported by the CMM workbench at three points. The experiment was carried out on a CMM 'ZEISS UMM500', whose technical specifications are given in Table 6.2. The developed model can be implemented on other CMMs as well and the accompanying software programmes. This is also evidenced by test measurements performed on the machine manufactured by DEA and software PC-DMIS.

TECHNISCHE UNIVERSITÄT WIEN	TU VIENNA	CNC-MESSABLAUF W-LAGE u. WERKSTÜCKAUFSPANNUNG	

WERKSTÜCK-NAME	FILE NR START / END	ZEICHNUNG	ARBEITSGANG
WORKING PART No.1.		DT-RP-01	LEVELIN (ZERO POINT)

TAS KOMB	TAS NR	ADR	PROG / AUFGABE	POS	ZUSÄTZLICHE HINWEISE
1	1	1	PLANE	A	4 POINTS
		2	DREHEN RAUM	W	3D COORDINATE TRANSFORMATION
		3	ZERO POINT	Z_0	Z=0
		4	PLANE	D	4 POINTS
		5	DREHEN EBENE	W	Z-AXIS ROTATION
		6	ZERO POINT	X_0	X=0
		7	POINT		1 POINT
		8	ZERO POINT	Y_0	Y=0

W-LAGE   WERKSTÜCKAUFSPANNUNG

Legend:
Plane: A, B, D, E
Cylinder: C1, C2
Cone: C3
Sphere: S1

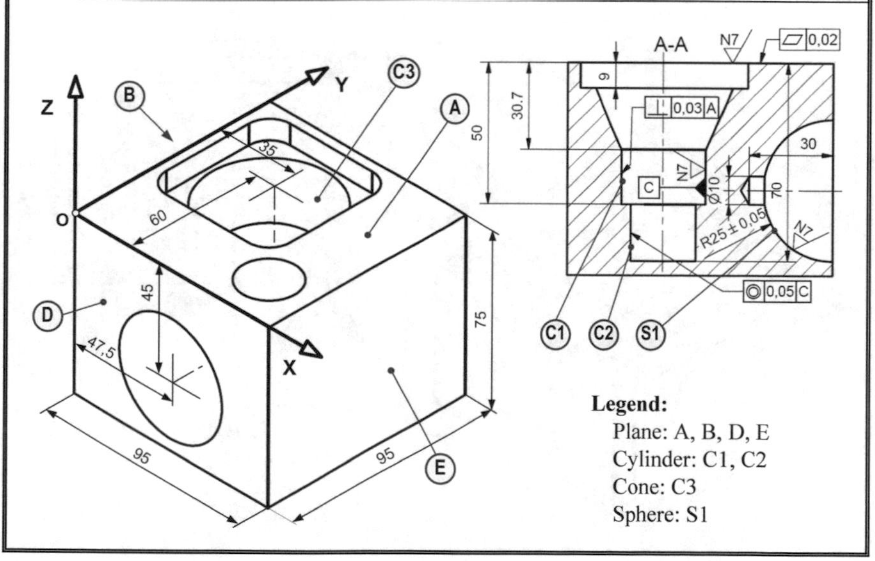

Fig. 6.10 Preparation for PMP 1 measurement

TECHNISCHE UNIVERSITÄT WIEN — TU VIENNA	**CNC-MESSABLAUF** *W-LAGE U. WERKSTÜCKAUFSPANNUNG*

WERKSTÜCK-NAME	*FILE NR START / END*	*ZEICHNUNG*	*ARBEITSGANG*
WORKING PART No.2.		*DT-RP-02*	*LEVELIN (ZERO POINT)*

TAS KOMB	*TAS NR*	*ADR*	*PROG / AUFGABE*	*POS*	*ZUSÄTZLICHE HINWEISE*
1	1	1	PLANE	A	4 POINTS
		2	*DREHEN RAUM*	W	3D COORDINATE TRANSFORMATION
		3	ZERO POINT	Z_0	Z=0
		4	PLANE	D	4 POINTS
		5	*DREHEN EBENE*	W	Z-AXIS ROTATION
		6	ZERO POINT	X_0	X=0
		7	POINT		1 POINT
		8	ZERO POINT	Y_0	Y=0

W-LAGE *WERKSTÜCKAUFSPANNUNG*

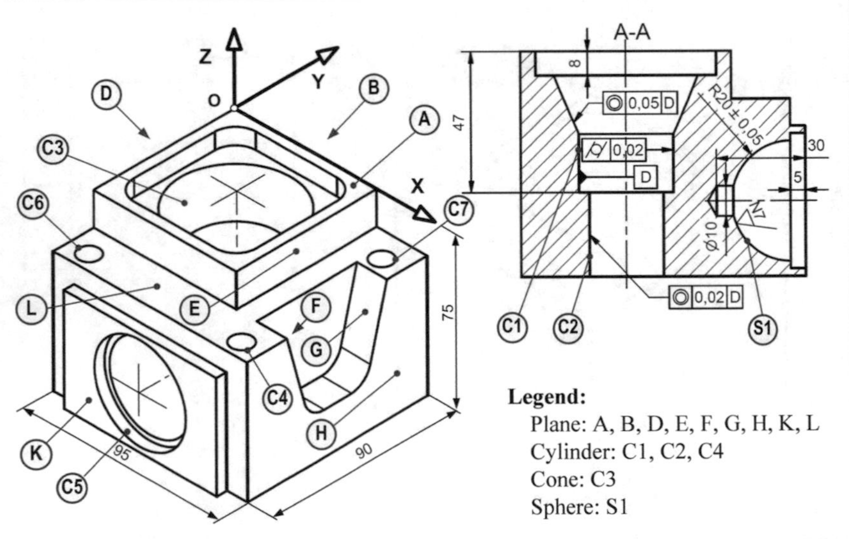
Z
Y
X
O
A
B
C3
C6
C7
D
E
F
G
H
K
L
C4
C5
95
90
75
A-A
8
47
R20 ± 0,05
30
5
Ø10
N7
0,05 D
0,02
D
0,02 D
C1
C2
S1
Legend:
Plane: A, B, D, E, F, G, H, K, L
Cylinder: C1, C2, C4
Cone: C3
Sphere: S1

Fig. 6.11 Peparation for PMP 2 measurement

Fig. 6.12 Experimental installation for the measurement of PMP1

Fig. 6.13 Experimental installation for the measurement of PMP 2 [1]

Table 6.2 Technical specifications for CMM

Number of axis:	3 (X, Y, Z)
Measuring range: X_{max} Y_{max} Z_{max}	500 mm 200 mm 300 mm
Automatic change of measurement sensor	Yes
Maximum weight of workpiece in kg	150
Resolution in μm	0.1
Software	ZEISS UMESS
MPE in μm	0.4 + L/600
Assurance y μm	0.2 μm

6.3   Measurement Results

Measurement results for both PMPs are given in Table 6.3 for all denoted primitives mentioned in the inspection plan for the parts. As it is evident from the tables, the measurement was repeated five times and standard deviation was calculated.

Results of the automatic inspection for two test PMPs indicate that all tolerances for the workpiece are within the limits prescribed by the drawing. This confirms that the described model is another successful approach to the automatic inspection for PMP and a solid basis for the development of intelligent approaches to inspection planning.

Complex geometry of PMP changes with a set of points, whose sequence defines a collision-free path. Representation of the measuring probe path by a set of points allows for its optimisation.

On the basis of the vector of a primitive and the fullness vector of a primitive, it is possible to automatically configure the measuring probes and thus reduce the preliminary measuring time.

The experiment included three types of measurements. The first measurement was conducted by online CMM programming, the second based on the programme defined by CAD software (Pro/ENGINEER) and the third according to the protocol defined by the model developed in this work. The results of comparison by applying the criterion of minimum distance travelled by a measuring probe, for individual primitives, are given in Table 6.4. In that case, as above mentioned, the needed quantities are designated as follows: d_s—length of a measuring probe path generated by software; d_m—length of a manually programmed measuring probe path; d_o—length of an optimised measuring probe path; I_1—enhancement of a path defined by the model relative to the software path; I_2—enhancement of a path defined by the model relative to the manually programmed path and I_3—enhancement of an optimised path relative to the initial path defined by the inspection planning model.

Measuring the inspection time is relatively complex and can be realised analogously to the application of RTM method in industrial robots. For this reason, in our research, the values of time were measured indirectly through the measuring probe

Table 6.3 Measurement results for test parts [1]

No.	Name	Label	Value in mm	1.	2.	3.	4.	5.	Deviation in μm
PMP 1									
1	Flatness	A	0.02	0.0005	0.0004	0.0005	0.0005	0.0004	0.1
2	Flatness	E	0.02	0.0001	0.0001	0.0002	0.0001	0.0001	0.0
3	Diameter	S1	±0.1	50.0851	50.0855	50.0852	50.0856	50.0855	0.2
4	X position	S1	±0.05	47.4611	47.461	47.4607	47.4611	47.4611	0.2
5	Z position	S1	±0.02	45.0148	45.0148	45.0146	45.0145	45.0146	0.1
6	Perpendicularity	A,C1	0.03	0.0014	0.0024	0.0023	0.0022	0.0023	0.4
7	Cylindricity	C1	0.04	0.009	0.0089	0.0091	0.009	0.009	0.1
8	Coaxiality	C1,C2	0.05	0.0474	0.0458	0.0456	0.0455	0.0462	0.8
9	Angle	C3	±0.5	39.2991	39.2991	39.2995	39.2982	39.2991	0.5
10	Parallelism	B,E	0.04	0.035	0.0346	0.035	0.0348	0.035	0.2
PMP 2									
1	Distance		±0.02	70.0111	70.0111	70.0112	70.0111	70.0106	0.2
2	Flatness	E	0.02	0.0072	0.0069	0.0071	0.0071	0.0072	0.1
3	Perpendicularity	A,C1	0.025	0.0143	0.0142	0.0142	0.0143	0.0148	0.3
4	Angle	A,G and A,F	0.2	15	15	15	15	15	0
5	Diameter	S1	±0.1	40.093	40.0936	40.0934	40.0936	40.0928	0.4
6	X position	S1	±0.05	47.5022	47.5026	47.5025	47.5021	47.501	0.6
7	Z position	S1	±0.3	49.7926	49.7926	49.7927	49.7925	49.7927	0.1
8	Parallelism	B,K	0.2	0.1366	0.1369	0.136	0.1361	0.1369	0.4
9	Cylindricity	C1	0.02	0.0042	0.0043	0.0041	0.004	0.0041	0.1
10	Angle	C3	±0.5	39.2986	39.2985	39.2983	39.2991	39.2992	0.4
11	Roundness	C5	0.03	0.0094	0.0092	0.0092	0.0101	0.0096	0.4

(continued)

Table 6.3 (continued)

No.	Tolerances		Value in mm	Measurement					Deviation in μm
	Name	Label		1.	2.	3.	4.	5.	
12	Position	C7,L	0.75	0.6175	0.6277	0.6332	0.6201	0.6125	8.2
13	Position	C6,L	0.75	0.5067	0.5021	0.5219	0.5073	0.5032	8
14	Position	C4,L	0.75	0.6471	0.6483	0.6416	0.6481	0.6424	3.2
15	Coaxiality	C1,C2	0.02	0.0068	0.008	0.008	0.0091	0.0092	1
16	Coaxiality	C1,C3	0.05	0.0409	0.041	0.0408	0.0406	0.053	5.4

Table 6.4 Results of comparison

Results	Features				
	Plane	Circle	Truncated hemisphere	Cylinder	Truncated cone
d_m (mm)	203.3896	154.3540	244.8584	290.9837	507.0366
d_s (mm)	202.6522	126.4417	183.6755	228.9870	440.1399
d_o (mm)	159.4604	110.2462	159.0962	172.2142	307.6091
D_{tot}	163.9298	119.1512	180.7477	179.7520	312.2379
$P_s = D_{tot}/d_s$ (%)	80.89	94.23	98.40	78.50	70.94
$P_m = D_{tot}/d_m$ (%)	80.60	77.19	73.82	61.77	61.58
$P_o = d_o/D_{tot}$ (%)	97.27	92.53	88.02	95.81	98.52
$I_1 = 100 - P_s$ (%)	19.11	5.77	1.6	21.50	29.06
$I_2 = 100 - P_m$ (%)	19.40	22.81	26.18	38.23	38.42
$I_3 = 100 - P_o$ (%)	2.73	7.47	11.98	4.19	1.48

path length. Given that the measuring probe path length is approximately directly proportional to measuring time, the values of probe path length approximately refer to the reduction of time required for measuring PMP. Therefore, the results given in Table 6.4 nearly apply to the reduction of measuring time.

6.4   Concluding Remarks

The development of intelligent systems for inspection planning is the imperative and prerequisite for the development of a new generation of technological systems and digital quality founded on a global interoperability model of the product. Interoperability model integrates CAD-CAM-CAI information in a digital environment and is a basis for virtual simulation and knowledge-based inspection planning for prismatic parts on CMMs.

The CMM is a basic element of flexible automation in production metrology and represents an essential factor in prismatic parts inspection. Today, the ongoing intensive investigations are related to solving the problem of intelligent inspection planning on a CMM, as a prerequisite for intelligent measuring machines development, on the one hand, and measuring time reduction given the increasingly expressed geometric and functional variability of products, on the other hand.

Inspection on measuring machines is based on complex software support for different classes of metrological tasks (tolerances). The performance of a uniform inspection plan on CMMs is a special problem that depends on metrological complexity of prismatic parts, intuition and experiential knowledge of the inspection planner. Elimination of intuition, knowledge representation, reuse and distribution through the development of an intelligent concept for inspection planning capable of decision-making at a given moment is the solution to the mentioned problem. In developing the intelligent concept, the emphasis is placed on generating optimal measuring probe path as a fundamental part of inspection for prismatic parts. In general, the concept involves:

- Development of ontological knowledge base through defining the entities and rules for browsing that base so as to prepare, connect and integrate geometric information with standard types of tolerances in an indirect way, by introducing metrological features and reducing them to geometric features,
- development of a global inspection plan which defines an optimal sequence of metrological primitives' inspection through analysis of the measuring probe accessibility and grouping of features according to the directions of approach,
- development of a local inspection plan that will generate the number and position of the measuring points, i.e. distribution for all metrological features and optimal measuring probe path for thus distributed points.

The actuality of planned research is reflected in several research themes related to:

- metrological interoperability,
- intelligent production metrology and,
- digital technology systems.

The relevance of initiating research and motivation is in the results achieved within the framework of research projects launched in the USA by the National Institute of Standards and Technology and leading European metrological laboratories, dealing with research of the development of intelligent concept of inspection planning for prismatic parts in order to reduce total measuring time on a CMM through reducing the component of time needed for inspection planning.

The aim of the whole research is to organise and efficiently utilise knowledge (experiential, intuitive) and deduction based on it, with the main purpose to perform a uniform inspection plan given relatively high metrological complexity and variability of product.

The crucial problem in inspection planning for prismatic parts is to connect tolerated measures of prismatic parts with basic geometric features according to geometric and dimensional tolerancing (G&T), product manufacturing information (PMI), integrated product information model (IPIM) and to bridge the gap between partially successfully developed two approaches to the development of CAIPP system, and these are geometric and tolerancing. Which of the primitives or group of features will be measured and when is also an open problem, whose solution establishes a relationship between ideal and real geometry of prismatic parts from ontological viewpoint. Research in this book is a contribution to the solution of above-mentioned two problems in order to accomplish long-expected concept of intelligent inspection planning for prismatic parts on a CMM.

Having in mind the above presented, the scope, domain and current directions of research, the following hypotheses that the research was originally based upon are confirmed:

- Engineering ontology was developed for the domain of coordinate measuring and knowledge base founded on it for defining the relationships between geometry and prescribed tolerances for prismatic parts.
- By modelling primitives for inspection, standard types of tolerances were reduced to basic geometric primitives and thus a direct relationship was established between tolerances and geometric primitives.
- A global inspection plan or a model of measuring probe path planning was developed.
- The existing Hammersley's method for measurement point distribution was extended, and modified formulas were obtained for measurement point distribution for basic geometric primitives and their parameters, taking into account three types of internal and external measuring surfaces such as flat, cylindrical convex and cylindrical concave.
- An algorithm was developed for optimal measuring probe path generation by defining collision zones and solving the travelling salesman problem applying the ant colony method.

According to the set and confirmed hypotheses, the scientific contributions of research in this dissertation are the ontological knowledge base, developed model of inspection planning, ant colony-based optimisation model and experimental verification of the planning and optimisation model.

On the basis of the analysis conducted about current state of methodologies for engineering ontology development, a method of engineering ontology development is proposed at the conceptual level in order to generate new methodology for engineering ontology development. Aside from knowledge reuse and distribution of one domain, the developed method defines ontology development for the needs of knowledge base construction as one of the basic components of the intelligent system for prismatic parts inspection on a CMM. By defining engineering ontology with the use of the presented method, a set of terms is defined, which mapped into the domain of knowledge base construction represents entities and relations between the entities.

The classes of engineering ontology are entities of a knowledge base, and relations between entities are properties of engineering ontology. Explicit application of the method is reuse and sharing of data as and logical structure of a knowledge base for intelligent inspection of prismatic parts on a CMM.

The result of the proposed method is an iterative process of ontology development for the domain of coordinate metrology in five steps. Method implementation in software Protégé was performed on the example of a measuring part, and it indicates that the presented approach to the engineering ontology development is fully justifiable for the domain of coordinate metrology and inspection of prismatic parts on a CMM.

Generation and application of the uniform inspection plan on a CMM represent a specific problem, which depends on metrological complexity of prismatic parts, inspection planner's intuition and background knowledge. Conducted research yields a knowledge base model for inspection planning in order to solve this problem and develop the intelligent system for inspection planning. Also, the result of this approach is defining affiliation of geometric features to individual types of tolerances through browsing a graph of knowledge base model. By graph browsing, general types of tolerances defined by a standard are linked to geometric features, so that it is possible to define metrological sequences and plan a measuring probe path.

Intelligent approach to inspection planning on CMMs is open research space, which, by integrating metrological and geometric complexity of prismatic parts, inspection planner's intuition and knowledge at output, produces a measuring protocol or control data list for a CMM. Research conducted in this doctoral dissertation gives the development of a new intelligent concept of inspection planning for prismatic parts in order to reduce measuring time and increase the autonomy of the inspection planning process based on automatic generation of the measuring protocol, measuring probe path optimisation by applying ant colony method, accessibility analysis, i.e. automatic measuring probes configuration. In general, research is directed to the development of the global and local inspection plan for prismatic parts on CMMs.

Experimental results indicate that the developed concept is one successful approach to intelligent inspection and a solid basis for further enhancement of autonomy extension. The concept integrates complex analyses contained in a developed model of inspection planning such as mathematical model for prismatic parts inspection, model of primitives for inspection, measurement point distribution, analysis of measuring probe accessibility, collision avoidance and measuring probe path planning. Simulation of the measuring process is based on the inspection planning model and enables visual inspection of the measuring probe path and generation of the measuring protocol. The application and benefit of the model are remarkable in the case of inspection for geometrically complex prismatic parts with a large number of tolerances and marked geometric variability in individual production.

Optimal measuring path is compared with online programmed measuring path and automatically generated measuring path in software Pro/ENGINEER (CMM module) for two specially built prismatic parts for research needs. The results of comparison between optimised path and online programmed path indicate at least 20% lower value for optimised path length compared to the Pro/ENGINEER path of at least 10% lower value for optimised path length, with identical setting of parameters. In addition to the presented concrete results, the advantage of ACO is simple implementation thanks to developed mathematical model that reduces measuring path to a set of points and optimisation problem to TSP.

Output from ACO is an optimised point-to-point measuring path for measuring basic geometric primitives. Analogously to programming machine tools using G-code, the obtained output can be virtually used for offline programming on a CMM by specifying optimised path in the point-to-point form and integrating into the measuring protocol for a concrete CMM.

Conducted research represents a response to industrial demands such as high geometric diversity and rapid placement of products on the market in order to reduce manufacturing lead time through reduction of inspection time and maintenance of a continuously required level of quality for inspection through automation of activities performed by inspection planner.

Limitation of the developed approach is its application only for inspection of prismatic parts not for parts with free-form surface, because the inspection planning model and the optimisation model are developed only for basic geometric primitives that prismatic parts consist of.

References

1. Stojadinovic S, Majstorovic V, Durakbasa N, Sibalija T (2016) Towards an intelligent approach for CMM inspection planning of prismatic parts. Measurement 92:326–339
2. ISO/TS 10360-2 (2009) Geometrical product specifications (GPS)—acceptance and reverification tests for coordinate measuring machines (CMM)—part 2: CMMs used for measuring linear dimensions
3. ISO/TS 15530-3 (2011) Geometrical product specifications (GPS)—coordinate measuring machines (CMM): technique for determining the uncertainty of measurement—part 3: use of calibrated workpieces or standards

Printed in the United States
By Bookmasters